Seabed Prehistory
Investigating the Palaeogeography and
Early Middle Palaeolithic Archaeology
in the Southern North Sea

By Louise Tizzard, Andrew Bicket and Dimitri De Loecker

Seabed Prehistory
Investigating the Palaeogeography and Early Middle Palaeolithic Archaeology in the Southern North Sea

By Louise Tizzard, Andrew Bicket and Dimitri De Loecker

Illustrations by
Kitty Foster and Karen Nichols

Wessex Archaeology Report 35
2015

Published 2015 by Wessex Archaeology Ltd
Portway House, Old Sarum Park, Salisbury, SP4 6EB
www.wessexarch.co.uk

Copyright © 2015 Wessex Archaeology Ltd
All rights reserved

British Library Cataloguing in Publication Data
A catalogue record for this book is available from the British Library

ISBN 978-1-874350-80-4

Designed and typeset by Kenneth Lymer
Edited by Jonathan Benjamin and Philippa Bradley
Cover design by Kitty Foster

Printed by Cambrian Printers

Front cover
Reconstruction of the MIS 8/7 floodplain landscape looking east

Back cover
Deployment of vibrocore © Wessex Archaeology and NOC
Multibeam bathymetry data from Area 240
Examples of hand axes and fauna recovered from Area 240 during dredging operations

Wessex Archaeology Ltd is a company limited by guarantee registered in England, company number 1712772. It is also a Charity registered in England and Wales number 287786, and in Scotland, Scottish Charity number SC042630.

Contents

List of Figures vi
List of Plates viii
List of Tables ix
Contributors ix
Acknowledgements x
Abbreviations xi
A note on dates, water depths and co-ordinate
    system conventions xi
Abstract xii
Foreign Language Summaries xiii

CHAPTER 1: INTRODUCTION
A fortuitous discovery 1
Project background 2
Study Area 3
Marine aggregate dredging and archaeology:
    managing the historic environment 5
The Aggregates Levy Sustainability Fund 6
Contribution to the wider field of
    submerged prehistory 8
Structure of this volume 9

CHAPTER 2: THE ORIGINAL FLINT AND FAUNAL ASSEMBLAGE
The flint assemblage 11
    Characterisation of the flake and
        core assemblage 12
    Characterisation of the hand axe assemblage . 14
    Raw material procurement 15
    Post-depositional artefact modifications 16
The Area 240 faunal remains: a summary 17
Discussion 17

CHAPTER 3: PALAEOGEOGRAPHIC RECONSTRUCTION METHODS
Introduction 21
Area 240 investigations 22
Stage 1: existing data review 22
Stage 2: geophysical data acquisition 28
    Positioning 29
    Single-beam echosounder data 29
    Sub-bottom profiler data 29
    Sidescan sonar data 31
    Magnetometer data 31
    Methodology discussion 31
    Comparison and integration of 2005 and
        2009 datasets 32
Stage 3: seabed sampling 32
Stage 4: palaeoenvironmental sampling 32
Stages 5 and 6: palaeoenvironmental assessment,
    analysis and dating 33

Subsequent work 36
Summary 36

CHAPTER 4: PREHISTORIC CHARACTERISATION OF AREA 240 AND THE SOUTHERN NORTH SEA REGION
Introduction 39
The present-day setting of the Palaeo-Yare 41
Area 240 palaeogeographic reconstruction 43
Pre-Yare Valley palaeogeography 45
    The Ur-Frisia delta plain: pre-Anglian
        (MIS 13 upwards; >478 ka) 45
    Extensive remodelling of the landscape:
        Anglian (MIS 12; 478–424 ka) 48
Early Development of the Palaeo-Yare 49
    A new drainage pattern: late Anglian
        (MIS 12; *c.* 434 ka) 49
    A marine incursion: Hoxnian (MIS 11;
        424–374 ka) 50
Channel and floodplain development during the
    Saalian (MIS 10–6; 374–130 ka) 51
    Overview 51
    Fluvio-glacial sediment deposition in
        the Palaeo-Yare 52
The last interglacial: Ipswichian (MIS 5e;
    130–115 ka) 56
Channel re-activation and continued development:
    Devensian (MIS 5d–MIS 2; 115 ka–11.7 ka) . . 57
    Early Devensian (MIS 5d–MIS 3;
        115 ka–54 ka) 58
    Mid- to late Devensian (MIS 3–MIS 2;
        54 ka–11.7 ka) 61
Early Holocene channel development and final
    transgression (<11.7 ka) 62
    Channel development 62
Post-transgression development 68
Preservation of sediments in Area 240 69

CHAPTER 5: THE CONTINUED SEARCH FOR ARCHAEOLOGICAL MATERIAL
Introduction 73
Remote sampling techniques 74
Grab sampling for artefacts 75
Area 240 seabed sampling survey 78
    Sampling strategy 78
    Positioning 78
    Video/stills photographs 78
    Two metre scientific trawl 80
    Grab sample acquisition and processing 81
    Transect 1 82
    Transect 2 85

    Transect 3 85
    Discussion 85
Dredging for archaeological, palaeontological and palaeoenvironmental material 86
    Sampling strategy 88
    Dredging vessel: assessment of dredge loads . . 88
    SBV Flushing Wharf 89
    Recovered flint 89
    Recovered palaeontological and environmental material 92
Summary 94

CHAPTER 6: EXAMINATION OF THE ARCHAEOLOGY, METHODOLOGICAL APPROACH AND MANAGEMENT OF AREA 240 AND FURTHER AFIELD

Introduction 95
The Area 240 assemblage formation and post-depositional modification 95
    The artefact assemblage 95
    Geological context of the artefacts 96
    Taphonomic considerations 97
    A local source of material and production? . . . 98
    Site formation scenarios 98
    Summary 100

The geographic and cultural setting of the Middle Palaeolithic assemblage within the Palaeo-Yare 100
    The potential for archaeological material in the aggregate block 100
    The significance of the artefact assemblage within the Palaeo-Yare 102
Faunal remains in Area 240 and the southern North Sea 103
    Geological context of the faunal remains . . . 103
The bigger picture 104
    Populating our palaeogeography 104
    Environment and resources 104
    Movement and colonisation 105
Method evaluation 106
    Reconstruction of the palaeogeography 106
    Sampling artefacts 107
    Summary 108
Management and mitigation 108
Further afield 109
Key conclusions 111

Bibliography 113
Appendix 1: original flint artefact descriptions . 125
Index 135

List of Figures

CHAPTER 1

Figure 1.1 Location of aggregate licence Area 240 from where the artefacts were dredged and SVB Flushing where the artefacts were discovered

Figure 1.2 Archaeological Exclusion Zone within Area 240

Figure 1.3 East Coast aggregate licence block with Area 240 highlighted

Figure 1.4 Generalised sea-level curve (based on stable oxygen isotope data as a proxy for eustatic sea level (data from Lisieki and Raymo 2005), glacial stages and archaeological periods referenced throughout this publication

Figure 1.5 Area of seabed subjected to palaeolandscape assessment including commercial and publically-funded projects

CHAPTER 2

Figure 2.1 The dredger tracks in Area 240 from which the archaeological material was recovered

Figure 2.2 Scatter diagram of maximum length against maximum width for flakes (dark green, n= 45), cores (light green, n= 8) and hand axes (blue, n= 31). The measurements for flakes are taken according to the 'axes' of the flakes (*cf*. Bordes 1961)

Figure 2.3 Graphical representation of the hand axe types (location of the maximum width versus the roundness of the edges) according to Bordes (1961)

Figure 2.4 Graphical representation of the hand axe types (location of the maximum width versus the elongation) according to Bordes (1961). Drawing after Debénath and Dibble (1994)

CHAPTER 3

Figure 3.1 Overview of geophysical and geotechnical data reviewed as part of the Stage 1 assessment

Figure 3.2 Illustration of gridding of data at: A) 50 x 50 m; B) 100 x 100 m; C) 200 x 200 m

Figure 3.3   Geophysical survey track plots acquired in 2009 for the boomer A), parametric sonar B), pinger C) and chirp D)
Figure 3.4   Vibrocore locations acquired and analysed in 2009
Figure 3.5   Palaeoenvironmental sampling of vibrocores VC2c, VC7c, VC8c1 and VC9c
Figure 3.6   Overview of marine geophysical and marine and terrestrial geotechnical datasets acquired for assessment of the Palaeo-Yare Valley

CHAPTER 4
Figure 4.1   Yare Valley location map with place names referenced in the text. Inset: Yare Valley drainage basin
Figure 4.2   Location of the offshore extents of the Breydon Formation (after Arthurton et al. 1994; Limpenny et al. 2011) and the location of the Cross Sand anomaly, the remains of probable chalk raft
Figure 4.3   Schematic illustrating the development of Area 240 and plan overview. Reference to the development stages are provided in the text
Figure 4.4   Seismic profile illustrating the late Anglian channel and Saalian (Unit 3a and 3b), early Devensian (Unit 4) infill sediments overlain by marine seabed sediments (Unit 8)
Figure 4.5   Generalised palaeogeography of the Middle Pleistocene illustrating the coastline and major rivers at approximately A) 1 MA and B) 750ka (after Cameron et al. 1992; Parfitt et al. 2010; Hijma et al. 2012)
Figure 4.6   Vibrocore log, depositional environment interpretation and photograph of VC2c (with OSL dates from VC3b) with a seismic section illustrating the targeted floodplain sediments
Figure 4.7   Limits of the Palaeo-Yare floodplain and relation to the late Anglian channel
Figure 4.8   Model of the base of late Anglian channel feature looking north-west illustrating the broad, shallow nature of the channel (vertical exaggeration: x15)
Figure 4.9   Generalised palaeogeography from the late Anglian (MIS 12) to the Saalian (MIS 6) (after Limpenny et al. 2011; Hijma et al. 2012)
Figure 4.10 MIS curve for the Saalian (MIS 10–6; 374–130 ka) illustrating OSL dating results for this period
Figure 4.11 Generalised palaeogeography during the Saalian (MIS 6) (after Murton and Murton 2012)
Figure 4.12 Vibrocore log, depositional environment interpretation and photograph of VC7c (OSL dates from VC7b) with a seismic section illustrating the targeted channel infill sediments
Figure 4.13 MIS curve for the Ipswichian (MIS 5e) and the Devensian (MIS 5d–2) illustrating OSL dating results for this period
Figure 4.14 Generalised palaeogeography between the Ipswichian interglacial (MIS 5e) and the late Devensian (MIS 2) (after Limpenny et al. 2011; Hijma et al. 2012)
Figure 4.15 Vibrocore log, depositional environment interpretation and photograph of GY_VC1 with a seismic section illustrating Saalian floodplain deposits (Unit 3b) and overlying early Devensian bank deposits (Unit 4)
Figure 4.16 Reconstruction of the environment and landscape of Area 240 during the early Devensian
Figure 4.17 Vibrocore log, depositional environment interpretation and photograph of VC9c (with OSL dates from VC9b) with a seismic section illustrating the targeted floodplain deposits (Unit 3b) overlain by mid-Devensian sediments (Unit 6) infilling shallow depressions
Figure 4.18 Location of early Holocene channel in relation to the late Anglian floodplain and channel. Core locations where sediments have been radiocarbon dated are illustrated
Figure 4.19 Radiocarbon dating results of a series of Unit 7 sediments
Figure 4.20 Vibrocore log, depositional environment interpretation and photograph of VC8c1 (with radiocarbon dates) with a seismic section illustrating the targeted Saalian floodplain deposits (Unit 3b) overlain by early Holocene channel deposits (Unit 7)
Figure 4.21 Multibeam bathymetry illustrating natural bedforms including underfilled channels and sandwaves up to 6 m high and the effects of dredging

CHAPTER 5
Figure 5.1   Archaeological finds in the southern North Sea reported through the protocol and A) Late Glacial antler point from Leman and Ower Bank (Godwin and Godwin 1933); B) large number of faunal remains recovered from Eurogeul shipping lane (Mol et al. 2006); C) Neanderthal skull fragment (Hublin et al. 2009)

Figure 5.2 Systematic sampling strategy used in the Palaeo-Arun and location of sample stations where worked flint were recovered
Figure 5.3 Proposed transects for seabed sampling
Figure 5.4 Transect 1 photograph locations at each sample station and photographs on the seabed taken by the ELVIS drop-down camera
Figure 5.5 Sample locations along Transect 1 targeting the floodplain deposits (Unit 3b) from which flint artefacts and faunal remains were recovered through clamshell grab sampling
Figure 5.6 Proposed transects and dredging lanes for monitoring of dredge loads
Figure 5.7 Illustrations of the hand axes recovered during monitoring of a dredge load. A) Find no. 1011; B) Find no. 1085; C) Find no. 1000

CHAPTER 6
Figure 6.1 Schematic illustrating the three site formation scenarios
Figure 6.2 Middle Palaeolithic context of the Area 240 assemblage in Northwest Europe *c.* MIS 9–7 (between around 300–200 ka)

List of Plates

CHAPTER 1
Plate 1.1 Photograph from the original collection of artefacts recovered from Area 240. The assemblage includes hand axes, flakes and cores
Plate 1.2 Outsize stockpile of aggregate at SVB Flushing

CHAPTER 2
Plate 2.1 Hand axe (X=Find number 240/14-01-08/078) found among the oversize gravel fraction at the SBV Flushing Wharf. Photograph by Jan Meulmeester
Plate 2.2 A. Elongated Levallois *sensu stricto* flake, which can be interpreted as a side scraper. B. Elongated extended Levallois flake. Scale 1:2. Photographs by Jan Meulmeester
Plate 2.3 A. Disc/discoidal core on a frost split piece of flint. B. Levallois *sensu stricto* core (nucleus Levallois récurrent). Scale 1:2. Photographs by Jan Meulmeester
Plate 2.4 A. Example of a cordiform hand axe. B. Example of a sub-cordiform hand axe. Scale 1:2. Photographs by Jan Meulmeester

CHAPTER 3
Plate 3.1 Deployment of surface-tow boomer sub-bottom profiler system
Plate 3.2 Deployment of chirp sub-bottom profiler system
Plate 3.3 On board recording of parametric sonar data
Plate 3.4 Deployment of vibrocore © Wessex Archaeology and National Oceanographic Centre
Plate 3.5 Geoarchaeological recording of a vibrocore
Plate 3.6 Assessing and counting plant remains

CHAPTER 4
Plate 4.1 Eroding cliff at Pakefield, Suffolk

CHAPTER 5
Plate 5.1 The Hamon grab deployed from the vessel to sample typically 10–15 litres of seabed sediment
Plate 5.2 Broken secondary flake recovered from clamshell grab sample during the East Coast Regional Environmental Characterisation survey
Plate 5.3 Deployment of the ELVIS drop-down camera
Plate 5.4 Recovery of the 2 m scientific trawl and sample from Transect 1. The sample comprises gravel, reworked peat and wood as well as seabed fauna
Plate 5.5 Recovery of the 280 litre hydraulic clamshell grab sample and on board processing of the sample through a 10 mm sieve.
Plate 5.6 Flint flake recovered during clamshell grab sampling (T1_G5)
Plate 5.7 Flint flake recovered during clamshell grab sampling (T1_G23)

Plate 5.8   Cervid/bovine centrotarsus recovered during clamshell grab sampling (T1_G5)
Plate 5.9   An archaeologist assesses dredged sediments for archaeology on board the *Arco Adur*
Plate 5.10 An archaeologist ready to assesses oversize material at the wharf

CHAPTER 6
Plate 6.1   Examples of artefacts illustrating the three taphonomic environments. A) minor modifications only; B) post-depositional surface modifications on one side; C) heavily weathered

List of Tables

CHAPTER 2
Table 2.1   Quantitative data on the flake tool typology
Table 2.2   Typological summary of the hand axes
Table 2.3   Radiocarbon dates of faunal remains recovered from Area 240 as part of the original discovery (courtesy of Jan Glimmerveen)

CHAPTER 3
Table 3.1   Comparison of 2005 and 2009 sub-bottom profiler and bathymetry datasets
Table 3.2   Details of sub-bottom profiler systems used in the Stage 2 geophysical survey

CHAPTER 4
Table 4.1   Sediment units identified in Area 240 with onshore and offshore equivalent formations
Table 4.2   OSL ages of channel and floodplain deposits (Unit 3b)

Table 4.3   Radiocarbon ages of Breydon Formation (Unit 7)

CHAPTER 5
Table 5.1   Summary of previous surveys using Hamon grabs for sampling archaeology
Table 5.2   Flint recovered from seabed sampling
Table 5.3   Sediment unit targeted on each transect
Table 5.4   Flint artefacts recovered during dredge monitoring
Table 5.5   Recovery of palaeontological and environmental material recovered during monitoring of dredge loads

CHAPTER 6
Table 6.1   Flint artefacts recovered from Area 240
Table 6.2   Key Middle Pleistocene Palaeolithic sites within the Palaeo-Yare catchment
Table 6.3   Faunal remains recovered from Area 240 between October 2010 and June 2012

Contributors

Dr Louise Tizzard
Wessex Archaeology
Portway House
Old Sarum Park
Salisbury
Wiltshire SP4 6EB
UK

Dr Andrew Bicket
WA Coastal & Marine
7/9 North St David Street
Edinburgh
Midlothian EH2 1AW
UK

Dr Dimitri De Loecker
Faculty of Archaeology
Leiden University
P.O. Box 9519
2300 RA Leiden
The Netherlands

Acknowledgements

This publication is primarily based on the project *Seabed Prehistory: Site Evaluation Techniques (Area 240)* which was funded by the Aggregate Levy Sustainability Fund and administered by English Heritage. The authors would like to thank Helen Keeley, Gareth Watkins, Peter Murphy, John Meadows, Jonathan Last and Gill Campbell of English Heritage for their assistance throughout the project. The authors would also like to thank Nigel Griffiths and Emma Beagley of Hanson Aggregates Marine Limited for their assistance throughout the Seabed Prehistory project and subsequent dredge and wharf monitoring work referred to in this publication.

Dr Ian Selby of The Crown Estate is thanked for his contributions, particularly concerning the dredging history and early discussions on the interpretation of Area 240. Our thanks are extended to all the captains, vessel crew and survey staff for their professionalism, assistance and support during the fieldwork surveys.

The project was managed on behalf of Wessex Archaeology by Dr Paul Baggaley and Dr Louise Tizzard. The original project design was developed by Dr Antony Firth. Fieldwork was undertaken by Dr Louise Tizzard (seabed sampling and geotechnical sampling), Dr Stephanie Arnott and Tina Michel (geophysics acquisition), John Russell (seabed sampling and geotechnical sampling), Kevin Stratford and Stuart Churchley (seabed sampling) and Patrick Dresch (geotechnical sampling). Assessment of geophysical data was undertaken by Dr Louise Tizzard, Dr Paul Baggaley, Patrick Dresch and Tina Michel. The assessments of plants and insects were carried out by Dr Chris Stevens and the mollusc assessment by Sarah Wyles. Pollen analyses were conducted by Dr Michael Grant and the ostracod and foraminifera analyses were carried out by John Russell. Finds conservation was provided by Lynn Wootten.

The radiocarbon dating was undertaken by Oxford Radiocarbon Accelerator Unit at Oxford University and the Scottish Universities Environmental Research Centre through the English Heritage Scientific dating team. Optically Stimulated Luminescence dating was carried out by Dr Philip Toms at the Geochronology Laboratory at the University of Gloucestershire. The diatom analysis was undertaken by Dr Nigel Cameron, University College London.

The flint artefacts recovered from SVB Flushing Wharf were assessed by Dr Dimitri De Loecker, a human origins specialist of the Faculty of Archaeology, University of Leiden. The authors would like to thank Mr Jan Meulmeester (Vlissingen, The Netherlands) for his intensive survey work and for offering the opportunity to study the Area 240 lithic collection. Moreover, we would like to express our gratitude for the time he invested in taking the numerous photographs. Thanks to Dr Veerle Rots (University of Liège, Belgium) for her succinct use-wear study and Mr Jan Glimmerveen (The Hague, The Netherlands) for his comments on the faunal remains. We are grateful to Mms Margot Kuitems (Leiden University, The Netherlands) for her work on the published artefact illustrations and to Dr Phil Glauberman (University of Connecticut, USA) for his work on Figures 2.3 and 2.4. Thanks also to Professor Raymond Corbey and Professor Wil Roebroeks (both Leiden University, The Netherlands) for their helpful comments on an earlier version of the lithic text.

Dredge and wharf monitoring was carried out in March 2011 by Dr Andrew Bicket and John McCarthy. Flint artefacts recovered during the Wessex Archaeology sampling survey in 2009 and dredge monitoring in 2011 were assessed by Dr Phil Harding.

The authors would like to thank Dr Paul Baggaley, Dr Michael Grant, Andrea Hamel, Dr Matt Leivers, Euan McNeill and John Russell for comments on early drafts of this volume. Illustrations were drawn by Kitty Foster and Karen Nichols.

Abbreviations

ADS	Archaeology Data Service	MV	Motor vessel
ALSF	Aggregate Levy Sustainability Fund	MALSF	Marine Aggregate Levy Sustainability Fund
AMS	Accelerator Mass Spectrometry	MAREA	Marine Aggregate Regional Environmental Assessment
BGS	British Geological Survey		
BMAPA	British Marine Aggregate Producers Association	MEPF	Marine Environment Protection Fund
CD	Chart Datum	MIRO	Mineral Industry Research Organisation
Cefas	Centre for Environment, Fisheries and Aquaculture Science	MIS	Marine Isotope Stage
COST	Cooperation in Science and Technology	MMO	Marine Management Organisation
		MTA	Mousterian of Acheulean Tradition
DCLG	Department for Communities and Local Government	NOC	National Oceanographic Centre
		nT	nanoTesla
Defra	Department for Environment, Food and Rural Affairs	OD	Ordnance Datum
		OS	Ordnance Survey
DGPS	Differential Global Positioning System	OSL	Optically Stimulated Luminescence
EH	English Heritage	REC	Regional Environmental Characterisation
EIA	Environmental Impact Assessment		
EMODNET	European Marine Observation and Data Network	RV	Research vessel
		SD	Standard deviation
EMS	Electronic Monitoring System	SLASHCOS	Submerged Prehistoric Landscape and Archaeology of the Continental Shelf
FAOL	Fugro Alluvial Offshore Limited		
HAML	Hanson Aggregate Marine Limited	SRTM	Shuttle Radar Topography Mission
Hz	Hertz	SUERC	Scottish Universities Environmental Research Centre
IAt	Acheulean Index		
IB	Biface Index	TWTT	Two-Way Travel Time
IF(s)	Index Facettage (stricte)	USBL	Ultra Short Baseline
IVS	Interactive Visualisation Systems		

A note on dates, water depths and co-ordinate system conventions

Throughout this volume calibrated radiocarbon dates are expressed in terms of cal BC. Dates produced by other scientific dating methods are presented in terms of ka. Dates over one million years ago are presented as 'Ma'. Ages of MIS boundaries are taken from Lisiecki and Raymo (2005).

Water depths are referred to as 'metres below (or above) Ordnance Datum (Newlyn)' abbreviated to 'm below (or above) OD'. Water depths surveyed for this project are also provided referenced to Chart Datum (CD) for Lowestoft and chart datum relative to OD (Newlyn) is -1.5 m (Admiralty Chart number 1543). Generalised references to change in global sea level are referred to as 'm below (or above) present-day' and are based on mean sea level.

Throughout the publication all drawings are presented in WGS84 UTM Zone 31 projection.

Abstract

The potential for Middle Palaeolithic sites to survive beneath the sea in northern latitudes has been established by intensive investigation within Area 240, a marine aggregate licence area situated in the North Sea, 11 km off the coast of Norfolk, England. The fortuitous discovery of Palaeolithic hand axes, Levallois flakes and cores, and other worked flint led to a major programme of fieldwork and analysis, funded by the Aggregate Levy Sustainability Fund and administered by English Heritage.

Between 7 December 2007 and 5 February 2008, 88 Palaeolithic artefacts and in excess of 100 fragments of faunal remains (including woolly mammoth, bison, horse and reindeer) were discovered by Mr Jan Meulmeester in stockpiles of gravel at the SBV Flushing Wharf. Based on the dates of discovery and through consultation with the licensee Hanson Aggregates Marine Limited it was established that the artefacts and faunal remains were dredged from a discrete locale within Area 240 in water depths of approximately 25 metres, and their provenance is judged to be secure.

The discovery and reporting of the Area 240 finds offered the rare opportunity to conduct a detailed study to establish the geological and geomorphological context of the recovered finds and to attempt to locate further artefacts.

A range of methodologies were tested to gauge their effectiveness in identifying and assessing sites of this type and produced numerous and diverse datasets – geophysical, geotechnical, palaeoenvironmental and archaeological – each subject to its own specialist methods of collection and analysis. Combined, the data has led to a comprehensive reconstruction of the development and preservation of the landscape and have furthered our knowledge on the quantity and extent of archaeological material within Area 240.

In addition to the initial discovery further artefacts recovered using seabed sampling techniques and during vessel and wharf monitoring of dredge loads indicate that the artefacts are not confined to a small, isolated zone of Area 240 but are more widespread.

The flint artefacts were recovered from floodplain deposits of the lower reaches of the Palaeo-Yare Valley, a fluvial system initially developed towards the end of the Anglian glaciation. The palaeogeographic reconstruction of Area 240 reveals a complex history of deposition and erosion. The preserved sediments derive from the time of the earliest known hominin occupation of Britain through to the Mesolithic, although not as a complete sequence.

The artefacts were primarily recovered from the Saalian floodplain sediments deposited in a cold, estuarine environment between 200,000 and 250,000 years ago. It is considered that the hand axes and Levallois products are contemporaneous in geological terms with taphonomically complex sedimentary contexts, as observed in several north-west European sites.

The Early Middle Palaeolithic assemblage from Area 240 has survived multiple phases of glaciation and marine transgression. The results have shown that submerged landscapes can contain preserved, *in situ* Palaeolithic artefacts. The investigations confirm that the artefacts are not a 'chance' find, but indicate clear relationships to submerged and buried landscapes that, although complex, can be examined in detail using a variety of existing fieldwork and analytical methods.

Zusammenfassung

Die Möglichkeit, dass mittelpaläolithische Fundstellen sich in nördlichen Breiten unter dem Meeresboden erhalten haben, konnte anhand intensiver Untersuchungen in Area 240 erbracht werden, einer Entnahmezone für marine Zuschlagstoffe, die 11 km vor der Küste der englischen Grafschaft Norfolk in der Nordsee liegt. Die zufällige Entdeckung von paläolithischen Faustkeilen, Levallois-Abschlägen und -Kernen sowie anderen bearbeiteten Flintgegenständen lieferte den Anlass für ein umfangreiches Programm archäologischer Untersuchungen und Analysen, das durch den Aggregate Levy Sustainability Fund gefördert und von English Heritage administrativ begleitet wurde.

Zwischen dem 7. Dezember 2007 und 5. Februar 2008 entdeckte Herr Jan Meulmeester 88 paläolithische Artefakte und mehr als 100 Tierknochenfragmente (u.a. Wollhaarmammut, Bison, Pferd und Rentier) in Kieshalden auf dem Gelände der SBV Aufspülfläche. Anhand des Entdeckungszeitraumes konnte in Zusammenwirken mit dem Konzessionsinhaber Hanson Aggregates Marine Limited ermittelt werden, dass die Artefakte und Tierknochen aus einem örtlich begrenzten Bereich in Area 240 aus einer Wassertiefe von ungefähr 25 m ausgebaggert wurden und ihre Herkunft somit als gesichert angesehen werden kann.

Die Entdeckung und Bearbeitung der Funde von Area 240 bot die seltene Gelegenheit, die geologischen und geomorphologischen Fundzusammenhänge der geborgenen Objekte in einer detaillierten Studie zu ermitteln und zu versuchen, weitere Artefakte zu lokalisieren.

Verschiedene methodologische Ansätze wurden hinsichtlich ihrer Effektivität bei der Prospektion und Beurteilung von Fundstellen dieses Typs getestet und lieferten zahlreiche, facettenreiche Datensätze (geophysikalische, geotechnische, paläobotanische und archäologische), jeweils mit den ihnen eigenen, speziellen Aufnahme- und Analysemethoden. Die Kombinierung dieser Datensätze hat eine umfangreiche Rekonstruktion der Entwicklung und Überlieferung dieser Landschaft ermöglicht und unsere Kenntnis von Menge und Ausmaß des archäologischen Materials im Bereich von Area 240 erweitert.

Neben der ursprünglichen Entdeckung ist aufgrund der Bergung weiteren Materials, das mithilfe von Probenentnahmen vom Meeresboden sowie im Rahmen von Maßnahme begleitenden Beobachtungen von Ausbaggerungsmaterial auf Schiffen und Spülflächen zutage kam, anzunehmen, dass die Funde nicht auf einen kleinen, isolierten Bereich innerhalb von Area 240 konzentriert, sondern weiträumiger verbreitet sind.

Die Flintartefakte wurden in den Schwemmsedimenten des Unterlaufs des Yare Urstromtals geborgen, eines gegen Ende der anglischen Vereisung (entspricht dem Elster Glazial) entstandenen Flusssystems. Durch die paläogeografische Rekonstruktion von Area 240 konnte eine komplexe Abfolge von Ablagerung und Erosion nachgewiesen werden. Die erhaltenen Sedimente decken der Zeitraum von der frühsten bekannten homininen Besiedlung Großbritanniens bis ins Mesolithikum ab, sie bilden jedoch keine komplette Abfolge.

Die Artefakte stammen hauptsächlich aus den Saale-zeitlichen Schwemmsedimenten, die unter kalten, ästuarinen Umweltbedingungen vor 200.000 bis 250.000 Jahren abgelagert wurden. Es wird davon ausgegangen, dass die Faustkeile und Levallois Artefakte in geologischem Sinne mit den taphonomisch komplexen Sedimentbefunden als zeitgleich anzusehen sind, wie dies auch in verschiedenen nordwesteuropäischen Fundstellen beobachtet wurde.

Der früh-mittelpaläolithische Fundkomplex von Area 240 hat mehrere Vereisungs- und Meerestransgressionsphasen überstanden. Die Ergebnisse haben gezeigt, dass sich in untermeerischen Landschaften paläolithische Artefakte *in situ* erhalten können. Darüber hinaus bestätigen sie, dass es sich nicht um Zufallsfunde handelt, sondern klare Verbindungen zu untermeerischen und fossilen Landschaften bestehen, die mit einer Auswahl vorhandener Feldforschungs- und Analysemethoden detailliert untersucht werden können.

Übersetzung: Jörn Schuster

Résumé

La possibilité, pour des sites du Paléolithique moyen, d'être conservés sous la mer dans les latitudes septentrionales, a été établie par une investigation intensive de l'Area 240, une zone de licence pour l'extraction d'agrégats marins située en Mer du Nord, à 11 km au large de la côte du Norfolk, en Angleterre. La découverte fortuite de bifaces paléolithiques, d'éclats et de nucléus Levallois, ainsi que d'autres silex taillés, a entraîné la mise en place d'un important programme de prospection et d'analyse archéologique, financé par l'Aggregate Levy Sustainability Fund et administré par English Heritage.

Entre le 7 décembre 2007 et le 5 février 2008, 88 artefacts paléolithiques et plus de 100 restes fauniques (comprenant du mammouth laineux, du bison, du cheval et du renne) ont été recueillis par M. Jan Meulmeester sur les tas de gravier de l'aire de stockage de matériaux SBV. En raison des dates de découverte et en consultant le titulaire de la licence, Hanson Aggregates Marine Ltd, il a été établi que les objets et restes fauniques ont été dragués sur une zone bien localisée de l'Area 240, sous env. 25 mètres d'eau: leur provenance est donc bien établie.

La découverte et l'analyse des trouvailles de l'Area 240 offraient l'opportunité rare de mener une étude détaillée afin d'établir le contexte géologique et géomorphologique des objets trouvées, et de tenter de localiser d'autres artefacts.

Différentes méthodes ont été testées pour évaluer leur efficacité dans l'identification et l'évaluation de ce type de site, produisant des bases de données nombreuses et variées – géophysiques, géotechniques, paléoenvironnementales et archéologiques – chaque spécialité ayant ses propres méthodes de collecte et d'analyse spécialisées. En définitive, les données ont permis une reconstruction complète du développement et de la préservation du paysage, et elles ont contribué à notre connaissance de la quantité et de l'étendue du matériel archéologique dans l'Area 240.

En plus de la découverte initiale, d'autres objets, récupérés grâce à des techniques d'échantillonnage du fond de mer, comme au cours du contrôle des chargement des dragues, en mer et sur les quais, indiquent que les objets ne sont pas confinés à une zone isolée de l'Area 240, mais qu'ils sont largement répandus.

Les objets de silex ont été récupérés dans les dépôts du lit majeur du cours inférieur de la paléovallée de la Yare, un système fluviatile initialement développé vers la fin de la glaciation Anglian (équiv. de la glaciation de l'Elster). La reconstitution paléogéographique de l'Area 240 révèle une histoire complexe de sédimentation et d'érosion. Les sédiments conservés datent de la première occupation d'hominines connus de la Grande-Bretagne jusqu'au Mésolithique, mais ils ne forment pas une séquence complète.

Les artefacts ont été récupérés principalement dans les sédiments des plaines d'inondation du Saalien, déposés il y a 200 000 et 250 000 ans environ, dans un environnement froid. On considère que les bifaces et les artefacts Levallois sont contemporains, en termes géologiques, des contextes sédimentaires de taphonomie complexe, comme on en a observé sur plusieurs sites du nord-ouest de l'Europe.

L'assemblage du Paléolithique moyen de l'Area 240 a survécu à plusieurs phases de glaciation et de transgression marine. Les résultats ont montré qu'il est possible de trouver des artefacts paléolithiques, préservé *in situ*, dans les paysages submergés. Les investigations confirment que les objets n'ont pas été découverts par «hasard», mais en relation étroite avec les paysages submergés et ensevelis qui, bien que complexes, peuvent être examinés de manière précise en utilisant les diverses méthodes d'exploration et d'analyse existantes.

Traduction: Jörn Schuster

Chapter 1
Introduction

A Fortuitous Discovery

This volume presents the results of a series of investigations of aggregate extraction 'Area 240', situated approximately 11 km off the coast of East Anglia, UK (Fig. 1.1). Between 7 December 2007 and 5 February 2008 gravel extracted from Area 240 yielded a series of important prehistoric discoveries: 88 worked flint artefacts (Pl. 1.1), including 33 hand axes, plus over 100 fragments of faunal remains including woolly mammoth, woolly rhinoceros, bison, reindeer and horse. The archaeological finds, datable by artefact typology to the Middle Palaeolithic (some 30,000 to 300,000 years ago), demonstrate the potential for such ancient material to survive beneath the sea. An appreciation of the rarity and importance of these artefacts and faunal remains, led to a major programme of fieldwork and analysis which began in autumn 2008 and continued until spring 2011.

This prehistoric cultural and faunal material was not actually encountered in the UK, despite its origin. Rather, the material was discovered in the aggregate outsize pile at SBV (Sorteerbedrijf Vlissingen) Flushing Wharf in Vlissingen, Netherlands (Pl. 1.2) by local archaeologist Jan Muelmeester, and facilitated by the local wharf manager Henk Strijdonk. Once it was established that the material had been dredged from Area 240, the aggregate area licensee, Hanson Aggregate Marine Limited, voluntarily placed a protective exclusion zone around that part of the seabed from which the cultural material was dredged (Fig. 1.2). Following their discovery, the artefacts were reported to English Heritage through the Marine Aggregate Industry *Protocol for Reporting Finds of Archaeological Interest* (BMAPA and English Heritage 2005) and the importance of the finds was subsequently acknowledged by the British Archaeology awards in 2008 as the Archaeological Discovery of the Year.

The discovery of the artefacts from Area 240 significantly changed archaeologists' understanding of the potential for prehistoric material in aggregate dredging areas in the North Sea. Although the potential for prehistoric material to be present was widely countenanced, and had been the subject of numerous investigations and research, actual

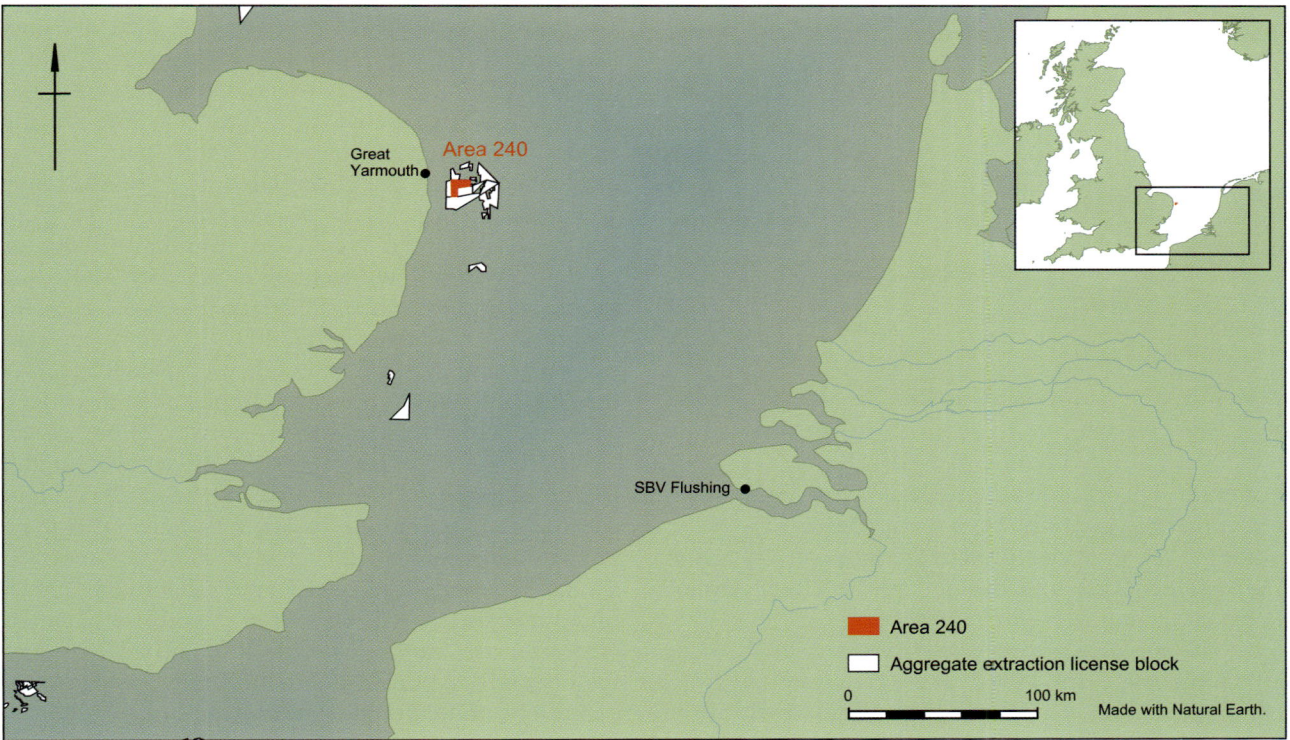

Figure 1.1 Location of aggregate licence Area 240 from where the artefacts were dredged and SVB Flushing where the artefacts were discovered

Plate 1.1 Photograph from the original collection of artefacts recovered from Area 240. The assemblage includes hand axes, flakes and cores

Plate 1.2 Outsize stockpile of aggregate at SVB Flushing

numbers of discovered artefacts were low. Also, attention has previously been directed primarily at the potential for archaeological material post-dating the Devensian glacial maximum (ie, of Late Upper Palaeolithic and Mesolithic periods).

Following the discovery, there was a need for both archaeologists and the aggregate industry to reconsider the assumptions upon which they managed the possible effects of aggregate dredging on the marine historic environment.

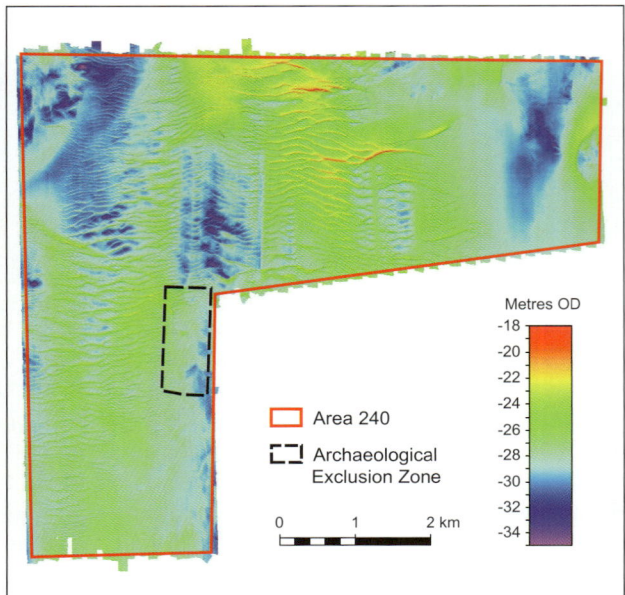

Figure 1.2 Archaeological Exclusion Zone within Area 240

Project Background

The discovery and reporting of the Area 240 finds offered the rare opportunity to conduct a detailed study to establish the geological and geomorphological context of the recovered finds and to attempt to locate further artefacts. Situated in water depths more than 25 metres below the surface and in an area with strong tidal currents and poor visibility, investigation of Area 240 proved challenging, and a carefully planned strategy was required.

Isolating the spatial provenance – the specific origin – of finds is a considerable task and its importance cannot be understated. For decades stone and bone artefacts, and faunal remains have been recovered from the North Sea through fishing or dredging but few have accurate location or contextual information (Godwin and Godwin 1933; Coles 1998; Hosfield 2007; Peeters *et al.* 2009). Even in the rare cases where spatial context has been recorded, such as the chance find in a vibrocore sample taken from deep water between Shetland and Norway (Long *et al.* 1986), further research on the geological context and spatial extent of the artefacts has not been carried out. Although the presence of isolated finds can provide a broad indication of the presence of artefacts in submerged areas (Hosfield and Chambers 2004), understanding the location and associated environmental context is critical for archaeologists to understand the nature of site formation, preservation conditions and cultural significance.

A multi-stage project was designed with an aim to be flexible and evidence-based; the development of each stage was designed in direct response to the results of the preceding stage. The principal aim of the project was to improve the future management of the effects (and potential effects) of aggregate dredging on the marine historic environment. There were three objectives to the project:

1. To refine practical techniques used to establish the presence or absence of prehistoric archaeological material (artefacts, deposits, faunal and other palaeoenvironmental material) on the seabed and to establish the character, date, extent, quality, preservation and special interest of such material, if present;
2. To improve the understanding of the character of the historic environment in the East Coast region, specifically its potential for prehistoric material;
3. To disseminate the knowledge gained throughout the scientific community, to industry, and the general public.

Between 2008 and 2011 a number of investigations were undertaken following two main strands of research (Objectives 1 and 2). The geological context of Area 240 was studied which included a detailed re-examination of geophysical and geotechnical data from industrial surveys, intensive geophysical survey of the area from which the artefacts and faunal remains were dredged, coring to obtain samples of the sedimentary sequence within Area 240 and accompanying palaeoenvironmental assessment and analysis, and the scientific dating of 12 samples. The second strand of investigation involved sampling the seabed for archaeological, palaeontological and palaeoenvironmental material using established techniques employed by the ecological and aggregate industries. The sampling exercise was conducted to assess the suitability for seabed sampling methods in order to observe this material, and their spatial distributions. Three sampling techniques were trialled: clamshell grabs, still photographic survey, and beam trawl and the ensuing results have led to a significant improvement toward our understanding of the character of the historic environment in the region. The results of each stage of the project (Wessex Archaeology 2009a–c; 2010a–b; 2011; De Loecker 2010) are available from the Archaeology Data Service (ADS) ALSF archive (archaeologydataservice.ac.uk).

This volume and associated publications (Tizzard *et al.* 2011; Tizzard *et al.* 2014) presents the project's results and aims to contextualise Area 240's contribution to prehistory, both geographically and archaeologically, in Britain and Northern Europe.

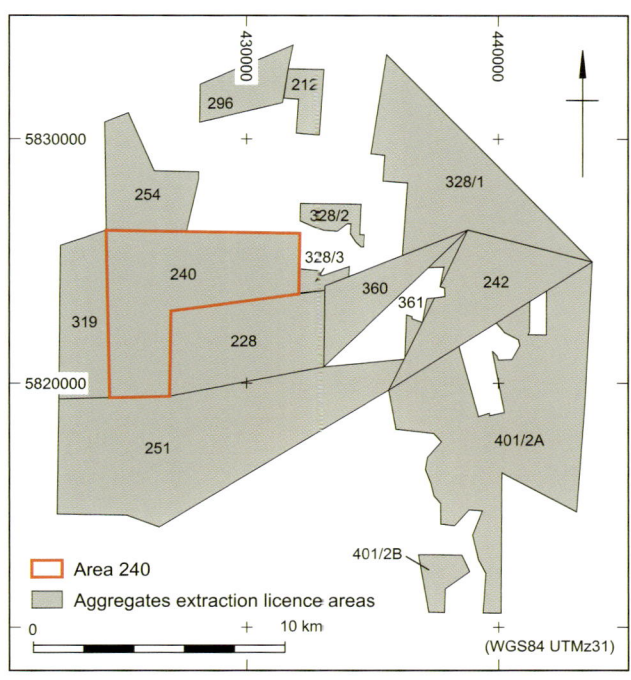

Figure 1.3 East Coast aggregate licence block with Area 240 highlighted

Subsequent to the conclusion of the project in spring 2011, work has continued in Area 240 and the surrounding aggregate block. In summer 2011, monitoring of the target dredge loads on the dredging vessel and at the wharf was carried out on behalf of Hanson Aggregate Marine Limited in order to further assess the presence or absence of archaeological material. The results of these trials are discussed in this volume. Further work has also been carried out on a wider scale on behalf of the aggregate industry. The aims were to extend the results from the palaeogeographic reconstruction of Area 240 into the surrounding aggregate block. Although this volume concentrates on the work carried out specifically in Area 240, the overarching results of the Palaeo-Yare catchment assessment have been incorporated where appropriate.

Study Area

Area 240 is one of a cluster of aggregate areas licenced to extract sand and gravel situated off the coast of Great Yarmouth, East Anglia (Fig. 1.3).

Marine aggregate dredging in the UK has a history stretching back to the 1700s when sand and gravel were used as ballast, shovelled by hand into the holds of barges beached on sand banks at low tide (Russell and Firth 2007). The industry in its modern form has its origins dating to the 1920s but it is since the 1960s that it has made an increasing contribution to UK

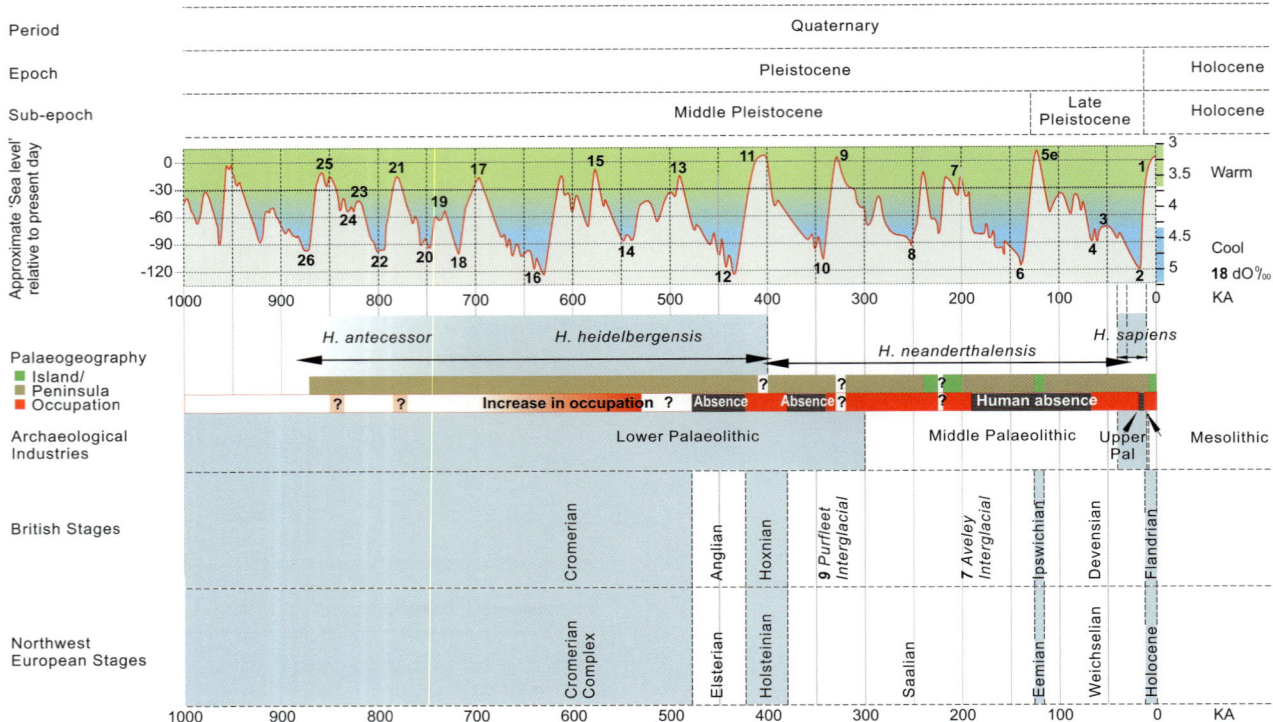

Figure 1.4 Generalised sea-level curve (based on stable oxygen isotope data as a proxy for eustatic sea level (data from Lisieki and Raymo 2005), glacial stages and archaeological periods referenced throughout this publication. The bold line on the sea-level curve indicates approximate depth of artefact recovery

aggregates supply. A system of nationally managed aggregate supply has been in operation since the 1970s to ensure the adherence to planning policies at local and regional levels.

The Crown Estate, which owns the UK's seabed from mean low water to the 12-mile territorial limit and holds the non-energy mineral rights out to 200 miles (as part of the hereditary possessions of the Sovereign, under *The Crown Estate Act* 1961), licences the marine aggregate industry by leasing 'Areas' for sand and gravel extraction. In 2012 the industry produced 16.8 million tonnes aggregate required for construction, coastal defence and beach replenishment. In 2012 the area of licenced dredging areas covers 711 km^2 of seabed, and while this may seem a considerable expanse, it accounts for a mere 0.147% of the UK Continental Shelf area (The Crown Estate and BMAPA 2013). Marine aggregates account for approximately 20% of the total aggregate supply in the UK (DCLG 2013).

Historically, the East Coast region has been an important source of marine aggregate with a total of 88,973 million tonnes of sand and gravel dredged between 1998 and 2008 (The Crown Estate and BMAPA 2009) and it is noteworthy that 90% of the regional dredging effort took place within an area of 16.4 km^2. Over recent years the volume of sediments and the area from which it is dredged has reduced as companies have released licence areas (or sub-areas) which reflect a change in emphasis by the licensees with increased efficient management of the remaining resources. Many of the dredging areas are reaching the end of their terms and are in the process of being renewed. The results of the work carried out in Area 240 have influenced the approach from industry with regards to mitigating the effects of dredging on the historic environment.

Area 240 is situated 11 km from the coast and the seabed is dominated by sandwaves up to 6 m high maintained by strong currents. Generally, water depths in the Area range between 20 and 35 m. However, for the majority of the last one million years, the area has been part of a coastal or inland environment as a result of lowered sea levels.

The Quaternary (Pleistocene and Holocene, Fig. 1.4) has been a period of fluctuating climate with corresponding oscillations in sea level. Sea-level change is complex and varies over spatial (local and global) and temporal (long-term and short-term) scales. During the Pleistocene three main processes influenced the development of the landscape, including Area 240 (Lowe and Walker 1997; Church *et al.* 2010):

- Global sea-level rise (and fall): glacio-eustatic adjustment resulting from ice formation/melt during alternating glacial and interglacial cycles;
- Land rebound (or subsidence): glacio-isostatic adjustment caused by changes in the weight of ice sheets (which further modified glacio-eustatic sea-level changes on a local scale);
- Tectonic uplift or subsidence, which produced smaller-scale fluctuations that would have had localised or perhaps regional consequences.

Chronologically, glacial and interglacial periods are broadly defined by their Marine Isotope Stages (MIS). This classification system comprises main stages and sub-stages, and while MIS stages reflect relatively long periods of time, it serves well as a coarse guide for past sea level and landscape change (Fig. 1.4).

During interglacial periods sea levels were relatively high, sometimes comparable to the present-day, whereas at the climax of glacial periods the sea levels fell to more than 120 m below present levels.

During the multiple cycles of transgression and regression (inundation or exposure of landscapes associated with sea-level change) various areas of the southern North Sea would have been repeatedly exposed. During these periods sediments would have been deposited and reworked, primarily through the repeated combination of fluvial action and glacial or permafrost melts. Marine transgression then followed, in some cases eroding the landscapes, and in other circumstances, inundating them in various states of preservation. Sediments found offshore reflect this pattern, although often the sequences are truncated, as existing sediments may have been completely or partially eroded before further deposition occurred. For a large part of the Middle and Late Pleistocene, rivers extended beyond the present-day shorelines onto the continental shelf. These extensions of existing rivers, enlarged by confluences that are now submerged, and swollen by glacial meltwater, would have been drainage systems of considerable size (Bridgland 2002). Area 240 is situated in the now-submerged lower reaches of the Palaeo-Yare river system; the onshore section includes the Rivers Yare, Wensum and Waveney. Reconstructing the depositional history of a coastal or submerged area requires an understanding of the seabed as an extension of the terrestrial landscape; these submerged terrestrial deposits are the target for aggregate extraction because of the nature of the material sought.

In order to understand the presence or potential for archaeology in now-submerged areas, the pattern of climatic deterioration and amelioration, and consequent periodic exposure and flooding of the continental shelf, needs to be compared with the known evidence for hominin presence and the (terrestrial) archaeological record (Fig. 1.4).

Marine Aggregate Dredging and Archaeology: Managing the Historic Environment

The results of the Area 240 work have and will continue to inform the process for aggregate licence application, as set out in the aims of the Area 240 seabed prehistory project.

The Crown Estate, as owners of the seabed, grants licences for aggregate dredging. Once a proposal for an extraction area is agreed in principle with The Crown Estate, the applicants must undertake an Environmental Impact Assessment (EIA) as part of its application in order to obtain a Marine Licence. This planning and consenting process for Licence Areas is the responsibility of the UK government via the Marine Management Organisation (MMO) in consultation with English Heritage as the primary advisor for the marine historic environment. EIAs have been required for aggregate dredging applications since 1989, under European Union Directive 85/338/EEC which stipulates that developers must undertake EIAs prior to project commencement. Once a Marine Licence has been granted, The Crown Estate issues the company applying for the licence with a production agreement.

EIAs ensure that the potential impacts of industrial activities are assessed for a wide range of environmental topics, including the historic environment (a term used by heritage managers to encompass archaeology and, in this case, prehistoric land surfaces and landscapes). There is potential for aggregate dredging operations to have significant impacts on archaeological material on the seabed. Site types include shipwrecks of various ages, aircraft (mainly military) crash sites and previously terrestrial prehistoric landscapes and associated cultural material which has been inundated by global sea-level fluctuations. Potential for prehistoric remains is often significant in areas that are specifically targeted by aggregate dredging companies. This is due to the presence of sand and gravels that were deposited by rivers during previous glacial periods when sea levels were lower and these areas were dry, habitable land. These types of locations, which would have been desirable for hominins, cause particular concern to archaeologists; the likelihood of sites being impacted by development therefore needs to be addressed by mitigation in the EIA process.

In the early years of the EIA process, there was a significant knowledge-gap regarding the marine archaeological record (ie, what was known and where it was located) and the actual archaeological remains that were extant on and under the seabed. This posed a challenge not only for archaeologists to identify threats to the fragile, non-renewable assets that make up the historic environment, but also for developers to minimise their impact on this un-quantified resource. The key was to determine whether important archaeological material existed in areas that also contained commercially-attractive aggregate deposits, and to develop methods that could enable dredging areas to be assessed and evaluated by archaeologists during the EIA process (Bicket et al. 2014). Additionally, because EIAs are generally localised, covering a relatively small area of the seabed, it was initially impossible to take the results from a single EIA and extrapolate them over a wider area. Only a wider, regional approach could provide valuable information about trends across the area (Firth 2006).

Although it was the process of EIA that provided some of the initial stimulus for the marine aggregate industry to address the implications of aggregate dredging on the historic environment, in the past decade the industry has gone far beyond the requirements of EIA to support the development of wider initiatives (Russell and Firth 2007). These initiatives have resulted in the development of industry guidance and the implementation of an innovative, industry-wide protocol for reporting finds of archaeological interest, and the funding provided for the ALSF has supported a wide range of projects that have considerably increased our knowledge of the historic environment (see Bicket 2011; Hamel 2011; Newell and Woodcock 2013).

In 2003, the British Marine Aggregate Producers Association and English Heritage produced a Guidance Note for assessing, evaluating, mitigating and monitoring the effects of marine aggregate dredging (BMAPA and English Heritage 2003). The guidance was targeted at marine aggregate developers, national and local curators and other contractors involved in the sector. The Guidance Note outlined the importance of both the marine historic environment and the aggregate industry, set out the regulatory framework, and identified possible effects of aggregate extraction on the historic environment. It also explained the role of the historic environment in the EIA process, and recommended mitigation measures including the implementation of exclusion zones, reducing impact by promptly seeking archaeological advice and remedying and offsetting. The Guidance Note also suggested the development of a finds protocol – to ensure that finds of archaeological interest, discovered during the dredging process, could be reported to the relevant archaeological curators (ie, English Heritage).

In August 2005 BMAPA and English Heritage introduced the Marine Aggregate Industry *Protocol for Reporting Finds of Archaeological Interest*, a protocol applicable to all BMAPA members, covering wharves, vessels and production licence areas (BMAPA and English Heritage 2005). Under the Protocol, archaeological material discovered in dredged loads or on the seabed by industry staff is reported to the Implementation Service through established channels.

The Protocol is supported and promoted to those working in the industry through an Awareness Programme which publishes a bi-annual newsletter, conducts site and wharf visits, produces and promotes web content and maintains vital communication channels, ensuring that industry stakeholders are empowered to protect and engage with the evidence of our shared marine heritage which is found during work in aggregate licence areas. Details of finds are disseminated by means of the newsletters and annual reports.

As of October 2014, there had been around 550 reports of finds submitted through the Protocol. Finds range from the Area 240 seabed prehistory project artefacts to wreckage from World War II aircraft and include an incredibly diverse range of archaeological material; individual reports vary from a single artefact to an entire collection of material.

The Aggregates Levy Sustainability Fund

The Area 240 project was funded through the ALSF and administered by English Heritage. This funding afforded the opportunity to study Area 240 in much greater detail than would have been appropriate through the EIA process. In turn, the results of the project and subsequent work have since fed into the EIA process for licence renewals.

In 2002, the UK Government imposed a levy on aggregate production to offset the environmental costs of extracting aggregate materials (Aggregates Levy Sustainability Fund – ALSF). That same year, the ALSF was set up so that a proportion of the revenue generated would be used for funding research aimed at minimising the effects of aggregate extraction on a wide range of environmental topics, including the historic environment. The ALSF provided the necessary funding to begin to answer many of the questions that had been raised through the EIA process. Initially set up as a two-year pilot project delivered by the Department for Environment, Food and Rural Affairs (Defra), the resounding success of the ALSF is demonstrated by the fact that the programme was extended three times, with the final projects completed in 2011. As part of the ALSF, the Marine ALSF (MALSF) focused on minimising the impacts of marine aggregate extraction. In 2004 the Marine Environment Protection Fund (MEPF) was established to commission MALSF projects. The MEPF was administered by the Centre for Environment, Fisheries & Aquaculture Science (Cefas) on behalf of Defra. MALSF/MEPF funds were provided to organisations through various distributing bodies, with English Heritage responsible for projects relating to the marine historic environment. An overview of the benefits of the MALSF was developed in *Marine Aggregates and Archaeology: a Golden Harvest* (Dellino-Musgrave *et al.* 2009) and also within an international context (Flatman and Doeser 2010). A comprehensive overview is also provided in Newell and Woodcock (2013). Projects funded by the MALSF have improved understanding of the historic environment resource and provided guidance, methodologies and further tools for assessing the resource and for mitigating against potential threats posed by aggregate extraction (Bicket 2011).

Four Regional Environmental Characterisation (REC) projects specifically examined submerged prehistory on the South Coast, East Coast, Outer Thames and Outer Humber regions (EMU Ltd 2009; James *et al.* 2010; Limpenny *et al.* 2011; Tappin *et al.* 2011), and provided an environmental baseline of high quality data over relatively large areas. Further projects such as the *North Sea Palaeolandscapes Project* (Gaffney *et al.* 2007) and *West Coast Palaeolandscapes Project* (Fitch *et al.* 2011) have provided regional-scale interpretation and assessments of archaeological potential of these landscapes. These large-scale studies, conducted at a lower resolution due to their scale, provide context for localised studies (Baggaley and Arnott 2013). In particular, the East Coast REC data provided regional context to the work carried out in Area 240.

Small-scale, high resolution surveys, such as those carried out as part of the project *Seabed Prehistory: Gauging the Effects of Marine Aggregate Dredging*, studied the archaeological potential of landscape features at a localised scale (Wessex Archaeology 2008a–f).

Between 2003 and 2008 Wessex Archaeology undertook a series of surveys with the aim of assessing the effects of dredging on aggregate resources (Wessex Archaeology 2008a–f). The projects were funded through the ALSF and administered in Round 1 by Mineral Industry Research Organisation (MIRO) and later, in Rounds 2 and 3, by English Heritage. The general aims of the *Seabed Prehistory* project were to better understand the extent and character of submerged prehistoric deposits and to develop methodologies for assessing and evaluating them in the course of aggregate area evaluation. In

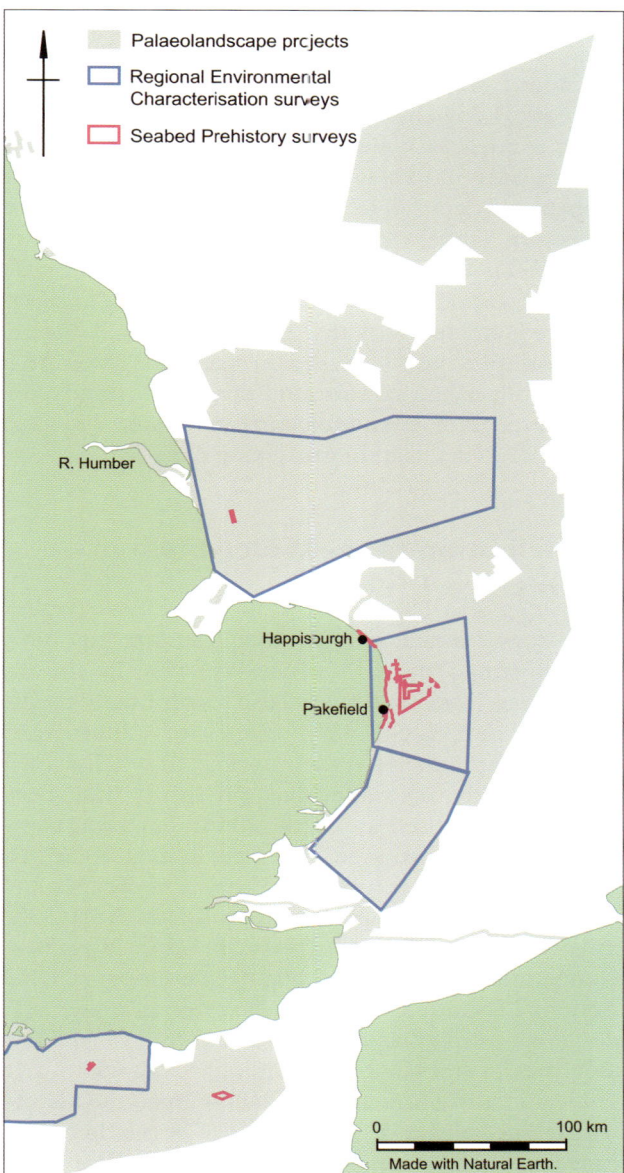

Figure 1.5 Area of seabed subjected to palaeolandscape assessment including commercial and publically-funded projects. Scale of assessments varies from the large-scale Regional Environmental Characterisation survey of the Humber, East Coast, Thames and South Coast to the local-scale surveys of the Seabed Prehistory project, surveys of the Humber, Happisburgh and Pakefield, Great Yarmouth, Eastern English Channel, and the Palaeo-Arun

order to achieve this, a series of surveys were undertaken which included acquisition of geophysical, geotechnical (vibrocore) and grab sampling data. The surveys were conducted in the North Sea and eastern English Channel and targeted a variety of landscape features deposited throughout different periods of the Pleistocene and Holocene.

The North Sea surveys (Fig. 1.5) were conducted off the mouth of the Humber Estuary and targeted a coastal palaeoenvironment deposited after the Last Glacial Maximum (*c.* 18,000 years ago). The Great Yarmouth study area investigated part of a wide submerged Palaeo-Yare Valley situated directly to the north of Area 240; these deposits date to late Middle Pleistocene (500 ka to 115 ka). Older, early Middle Pleistocene deposits (*c.* 700 ka) were the focus of the nearshore surveys at Happisburgh and Pakefield, situated on the East Anglian coast. In the English Channel, the offshore extension of the River Arun was surveyed and part of the eastern English Channel river complex was surveyed in aggregate Area 464. The results of the surveys, and more importantly, the development of methodological approach, provided the foundations used to investigate Area 240.

Contribution to the Wider Field of Submerged Prehistory

The work carried out in Area 240 has not only contributed to the EIA process specifically for the East Coast region but has also contributed to the wider multi-disciplinary field of submerged prehistory.

The study of submerged prehistory has been formally recognised as an important, emerging topic on a pan-European scale (see Benjamin *et al.* 2011). In 2009, the Submerged Prehistoric Landscapes and Archaeology of the Continental Shelf (SPLASHCOS) initiative was funded through the European Cooperation in Science and Technology (COST Action TD-0902). The four-year programme was designed, within the parameters of COST, to facilitate meetings, workshops, training activities and publications on the topic of submerged prehistory with a special focus on early stage researchers and capacity building for a new generation. The main objectives of SPLASHCOS focused on promoting research on the investigation, interpretation and management of the drowned landscapes and prehistoric archaeology of the European Continental Shelf, to create a structure for the development of new interdisciplinary and international research proposals, and to provide guidance to heritage professionals, government agencies, commercial organisations, policy makers and a wider public on the relevance of underwater research to a deeper understanding of European history, reconstructions of palaeoclimate and sea-level change, and the social relevance and likely future impact of these changes (Bailey and Sakellariou 2012). The inception of the European Marine Board's working group Submerged landscapes and the underwater cultural heritage (SUBLAND) represents the continued development and promotion of such research. As the disciplines of marine and environmental sciences meet with archaeology and sea-level studies, research topics and methods are advancing and the variety of results from research into the submerged past is striking.

Further research frameworks published over recent years have had a more regional focus. In 2009 the North Sea Prehistory Research and Management Framework (NSPRMF) was published with the purpose:

> ...to facilitate the large-scale systematic and interdisciplinary study and preservation (where possible) of a unique sedimentary and archaeological record of some two million years that is currently submerged beneath the waters of the southern North Sea (Peeters *et al.* 2009, 7).

A number of key themes were proposed addressing the various aspects of the human use of former landscapes in the southern North Sea Basin from spatially and temporally variable perspectives. Themes that are appropriate for the work carried out at Area 240 include: stratigraphical and chronological frameworks, palaeography and environment, and Pleistocene hominin colonisations of northern Europe. The Area 240 research also contributes to a number of themes outlined in *People and the Sea: A Maritime Archaeological Research Agenda for England* which focuses on maritime and the marine historic environment research agendas for England (Ransley *et al.* 2013).

The development of palaeoenvironmental and prehistoric sea-level rise studies (eg, Flemming 1969; Shackelton *et al.* 1984; Dawson 1984; Pirazzoli 1985; 1996; Fairbanks 1989; van Andel 1989; 1990; Christensen 1995; Lambeck *et al.* 1998; 2001; Shennan and Horton 2002; Lisiecki and Raymo 2005; Kopp *et al.* 2009; Dutton and Lambeck 2012) has significantly influenced and directly contributed to the interdisciplinary study of submerged prehistoric archaeology worldwide. The processes involved in the formation of, and the scope for encountering and understanding, prehistoric archaeology on submerged landscapes has been introduced or summarised in various publications (eg, Masters and Flemming 1983; Johnson and Stright 1992; Fischer 1995; Flemming 1998; Bailey and Flemming 2008; Benjamin *et al.* 2011; and specific to Britain and the southern North Sea: Coles 1998; Flemming 2004; Momber *et al.* 2011) and the use of geophysical and geotechnical techniques, integrated with palaeoenvironmental analyses, have provided the means to reconstruct and understand these landscapes (eg, Gaffney *et al.* 2007; Mellett *et al.* 2012). It is now clear that submerged palaeolandscapes can fundamentally enhance our knowledge of the human past, and submerged prehistory has been identified as a broad priority for archaeological research (*cf.* Bailey 2004). Considering the number of important sites throughout Europe to have been located and studied by archaeologists, it has become obvious that the modern coastlines, which were not the boundaries in prehistory, should not be treated as such today.

The data acquired during the course of the project have also contributed to a large volume of data within UK waters that have been used to assess submerged palaeolandscape features. A recent English Heritage funded project *Audit of Current State of Knowledge of Submerged Palaeolandscapes and Sites* (Wessex Archaeology 2013a) compiled a database of all known surveys, including government-funded and commercial work, in English waters. The results indicate large areas of the North Sea (Fig. 1.5), Eastern English Channel and Irish Sea have been surveyed, unsurprisingly, most densely in areas where there is a high level of commercial activity (such as proposed windfarm and aggregate extraction areas).

These projects focus on palaeolandscapes and the potential for archaeology therein. Although the Area 240 project is localised and only accounts for a small percentage of that landscape coverage, the work carried out is based on the recovery of artefacts. Such, assemblages in the submerged environment, at present, are very rare. As such, the work carried out at Area 240 has significantly contributed to our understanding of North Sea prehistory and how we study submerged heritage. The following chapters illustrate why we think this to be so.

Structure of this volume

Chapter 2: The Original Flint and Faunal Assemblage sets out a detailed description of how the initial discovery was made and analysis of the artefacts and faunal remains.

Chapter 3: Palaeogeographic Reconstruction Methods discusses the techniques used for investigating the site and the research, which involved the adaption of ecological sampling methods (photography, beam trawling and clamshell grab sampling), to establish the presence or absence of archaeological material (artefacts, deposits, faunal and other palaeo-environmental material).

Chapter 4: Prehistoric Characterisation of Area 240 and the southern North Sea Region presents the results of the palaeogeographic reconstruction in Area 240 and discusses the findings in relation to the development of the wider southern North Sea.

Chapter 5: The Continued Search for Archaeological Material presents the rationale and results from a variety of sampling campaigns aimed at consolidating the provenance and environmental context of the artefacts.

Chapter 6: Examination of the Archaeology, Methodological Approach and Management of Area 240 and Further Afield presents an overview of the artefact assemblage and the associated geological context and provides an evaluation of the methodologies used. The chapter also reviews the management and mitigation strategies employed in Area 240 and assesses the wider impacts of the discovery of the site at Area 240.

Chapter 2
The Original Flint and Faunal Assemblage

The original Palaeolithic artefacts and faunal remains were recovered from stockpiles of gravel at the SBV Flushing Wharf in Vlissingen, Netherlands between 7 December 2007 and 18 March 2008 by local archaeologist Jan Meulmeester (Fig. 1.1). Based on the dredging vessels recorded Electronic Monitoring System (EMS) data provided by the licensee Hanson Aggregate Marine Limited, it was determined that the material was dredged between 7 December 2007 and 5 February 2008. It was then possible to attribute the finds to an area of the seabed measuring approximately 3.5 x 1 km, referred to as the Archaeological Recovery Zone (ARZ), on the eastern side of licence Area 240. Since 1993 it has been compulsory that all vessels dredging on licence areas in UK waters should be fitted with an EMS which automatically records the date, time and position of all dredging activity and every month this information is supplied to The Crown Estate (The Crown Estate and BMAPA 2010). The data allowed discrimination between the vessel's movements and timing of dredging activity which, in turn, allowed for an assessment on the provenance of the archaeological and palaeontological material. Subsequent to the reporting of the discovery, Hanson Aggregate Marine Limited stopped dredging in the area and put in place an Archaeological Exclusion Zone (AEZ). This zone measured approximately 1.5 x 0.6 km and was located to the southern part of the ARZ, where it was deemed most likely that the recovery had been made (Fig. 2.1).

In the context of this discussion, the term *assemblage* is consciously defined to refer to all of the artefacts found in the Area, rather than the more specific use of the term to refer to an associated set of contemporary artefacts found in a single unit (Darvill 2008, 27). Although the collection of material appears to have more than a single context and associated taphonomic process(es), given the nature of the exploration and for the sake of organisational simplicity, it is considered it appropriate to refer to the material as an assemblage in the broadest sense. An assessment of the recovered flint assemblage was independently carried out by Dr Dimitri De Loecker, human origins specialist at the faculty of Archaeology, Leiden University (De Loecker 2010). Faunal remains amounting to approximately 130 fragments were recovered along with the flint artefacts and underwent analysis by Mr Jan Glimmerveen in the Netherlands. The results of these works are presented in this chapter.

The Flint Assemblage

The original archaeological assemblage amounts to 88 flint artefacts which have been classified as 33 hand axes, eight cores, and 47 complete and fragmentary flakes (see Appendix 1). On 13 of the flakes retouch was evident. The majority of the artefacts indicate that rolled raw flint nodules were likely to have been sourced from river deposits, such as exposed gravel bars or terraces. The condition of the artefacts indicates that several taphonomic processes were responsible for the formation of the assemblage.

A detailed attribute analysis, initially developed for the Dutch Middle Palaeolithic Maastricht-Belvédère 'scatters and patches' (De Loecker and Schlanger 2006), was adapted for assessing the Area 240 flakes and cores. A system based on recorded measurements and ratios, developed by Bordes (1961; Debénath and Dibble 1994, 132), provided a classification of the different hand axe groups (De Loecker 2010). What follows is a brief typo-/technological and taphonomic

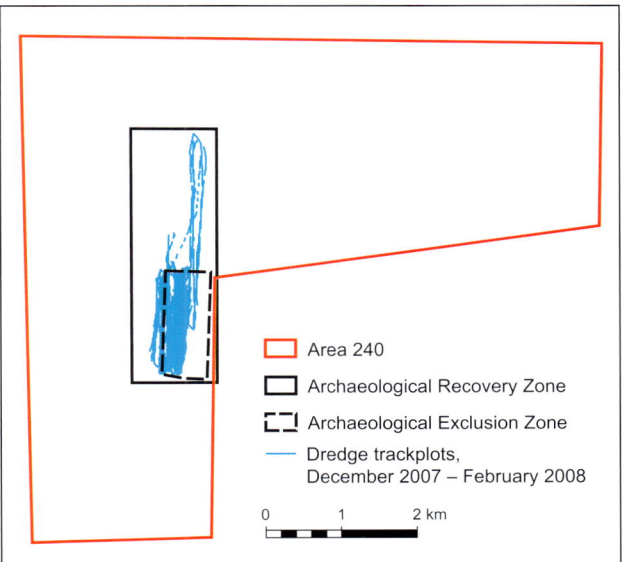

Figure 2.1 The dredger tracks in Area 240 from which the archaeological material was recovered. The majority of the tracks are located within the Archaeological Exclusion Zone

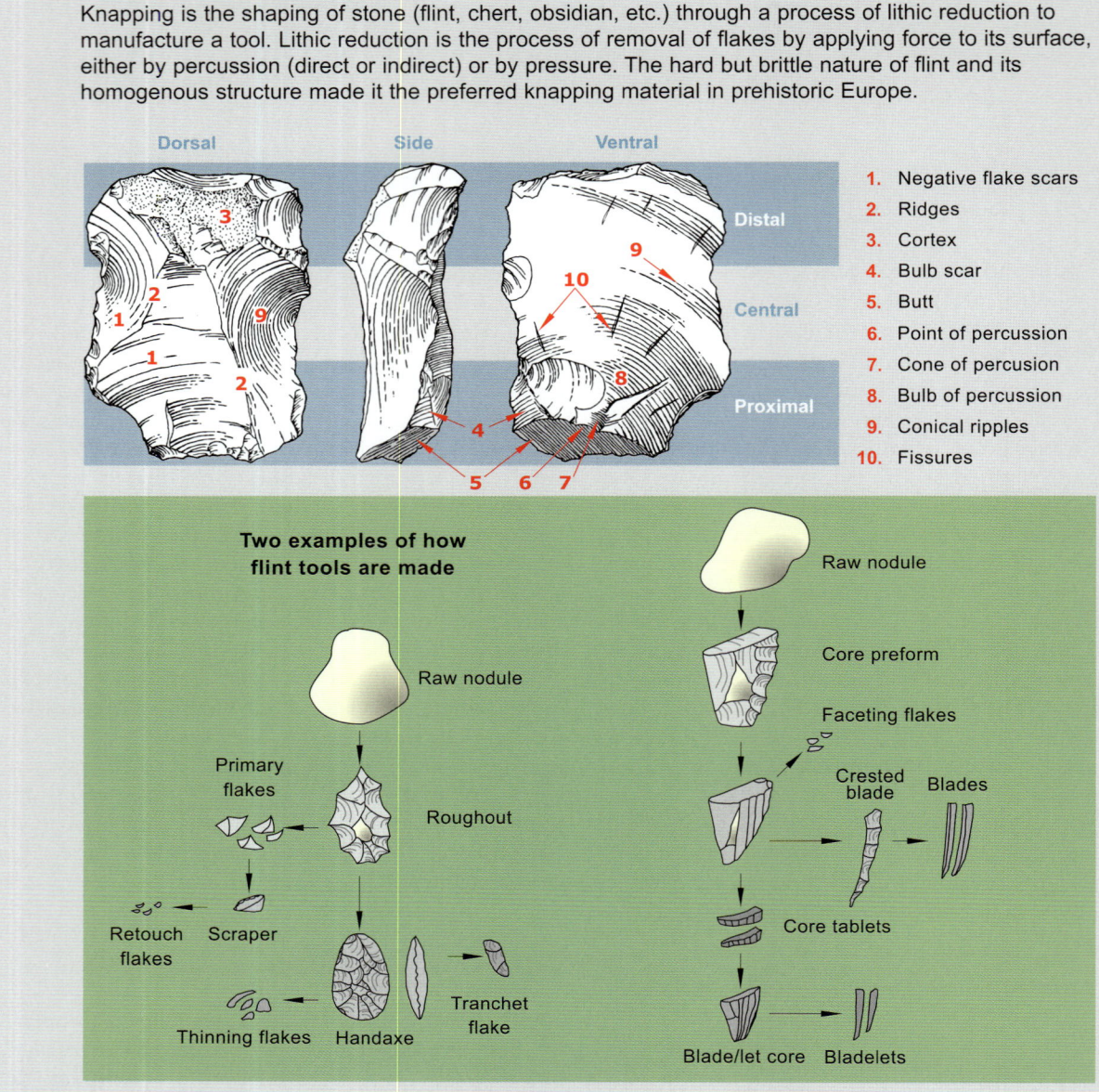

overview of the lithic material followed by a discussion on the raw material procurement, post-depositional conditions and production of flakes and cores and tool (specifically hand axe) manufacture. Two hand axes were not available for analysis so the following results are based on 86 flint artefacts.

Characterisation of the Flake and Core Assemblage

The Area 240 assemblage consists of 47 large and thick flakes, which show a technological heterogeneity. Strikingly, the sample contains a large diversity of both very simple flakes (ie, flakes, produced one after the other, from the same striking platform, in the same direction and without preparation) as well as well-prepared flakes which are produced in an operational scheme where the core was initially prepared before a flake was removed. Maximum dimensions less than 100 mm are clearly underrepresented with only two flakes less than 90 mm (artefacts 240/03-03-08/065 and 240/14-03-08/082). Flakes present are somewhat longer than wide and undoubtedly reflects a biased assemblage based on size (Fig. 2.2) as a consequence of the processing procedures at the SBV Flushing Wharf. At Flushing Wharf when the aggregate is offloaded from the dredging vessel it is stored in stockpiles and as each stockpile is processed the aggregate is graded

Plate 2.1 Hand axe (X = Find number 240/14-01-08/078) found among the oversize gravel fraction at the SBV Flushing Wharf. Photograph by Jan Muelmeester

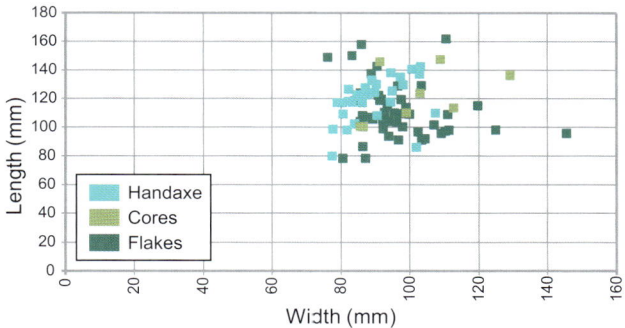

Figure 2.2 Scatter diagram of maximum length against maximum width for flakes (dark green, n= 45), cores (light green, n= 8) and hand axes (blue, n= 31). The measurements for flakes are taken according to the 'axes' of the flakes (cf. Bordes 1961)

according to size classifications. Oversize material (>63 mm) is delivered to a temporary stockpile and it was from here that the flint artefacts and faunal remains were recovered. As such, potential smaller lithic artefacts (such as flakes and debitage less than 63 mm) are excluded, inaccessible for discovery alongside larger finds. Additionally, the oversize material stockpile itself was a large mound (Pl. 1.2) and visually inspecting such a mound for the purpose of identifying worked flint requires time and an experienced eye. Under these circumstances, it is natural that larger, recognisable, or more obvious artefacts are recognised first (Pl. 2.1).

The Area 240 assemblage is characterised by few (only four) retouched or facetted butts. Flake scars from earlier stages in the reduction process are generally used as a striking platform (ie, plain and dihedral butts). In contrast to the scarcity of preparation on the striking platforms (or butts), the dorsal flake surfaces are generally well-prepared. The majority of the flakes display a complex dorsal pattern, in which previous flakes were struck from several directions. Centripetal or radial dorsal patterns predominate.

Moreover, 20 flakes were interpreted as Levallois *sensu stricto*, or as extended Levallois products, that is to say the products resemble classic Levallois but only display a portion of the technological features (*cf*. Bordes 1961; Boëda 1984; 1986; 1988; 1994; Van Peer 1992) (Pl. 2.2). An oval flake shape is commonly found on Levallois *sensu stricto* flakes, while feather terminations dominate. The assemblage also shows that some flakes are overstruck or display hinge terminations; the accidental result of too forceful trimming blows, struck in the same direction as the major flake removals. Hard-hammer percussion is predominant.

Notably, 13 (possible) flake tools are present and most can be described as Levallois *sensu stricto* (Table 2.1). Amongst these pieces is a large single

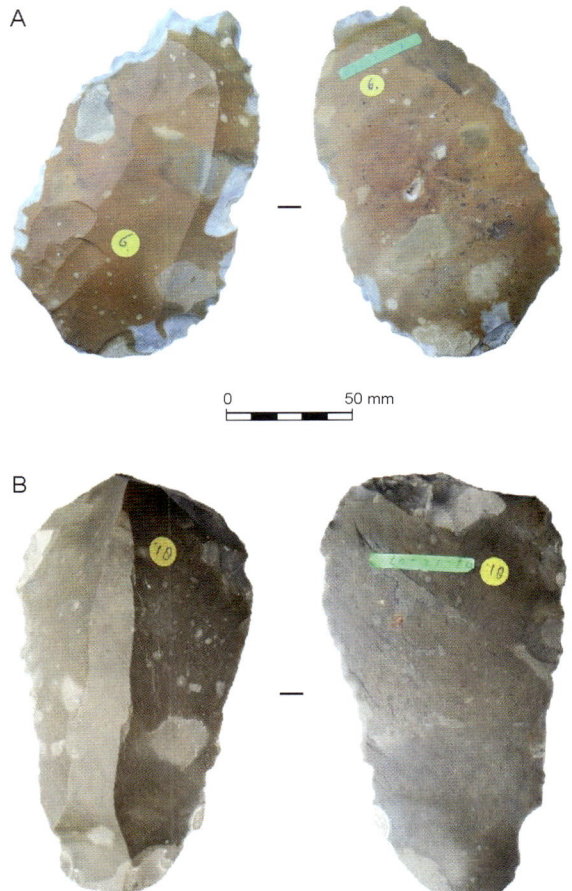

Plate 2.2 A) Elongated Levallois sensu stricto flake, which can be interpreted as a side scraper. B) Elongated extended Levallois flake. Scale 1:2. Photographs by Jan Meulmeester

Plate 2.3 A) *Disc/discoidal core on a frost split piece of flint*. B) **Levallois** sensu stricto *core (nucleus Levallois récurrent)*. Scale 1:2. Photographs by Jan Meulmeester

convex side scraper which was produced on a centripetal 'elongated' Levallois *sensu stricto* flake (Find number 240/18-03-08/060). The 'fresh' looking artefact was briefly examined on microscopic use-wear traces by Dr Veerle Rots, though no use-wear was detected, as the implement exhibited too much post-depositional edge damage (Rots pers. comm.).

Four of the eight cores can be classified as 'disc/discoidal' (Pl. 2.3, A), and three are classified as extended or as *sensu stricto* Levallois, (Pl. 2.3, B). One core (240/11-02-08/046) is interpreted as flaked-flake (*cf.* Ashton *et al.* 1992). This indicates that the assemblage is characterised by a prepared core approach, with several 'classic' Levallois cores. All cores have a maximum length between 100 and 149 mm, and the average length (122.2 mm, σ = 19.1 mm) is about 20 mm longer than the average width (102.0 mm, σ = 14.8 mm). Core measurements demonstrate that rather large and wide, but 'thin' nuclei were discarded (Fig. 2.2). This could suggest that that the cores were mainly reduced from one striking surface. In case of a Levallois situation this could mean that the core-reduction strategy was more oriented towards the production of single 'desired products' (*éclats préférentiels*) from a single striking surface. If this is the case, than these nuclei are just discarded exhausted cores. Alternatively, the core measurements could relate to the morphology of the raw material nodules. All cores show cortex, which on six of the cores covers more than a quarter of the total surface area. This indicates that nodules were introduced at a location with minimal decortication and that not all faces of the cores were exploited. This indicates that several flint nodules were selected without any (or hardly any) preparation or testing. Subsequently the flint blocks were probably mainly reduced from "one" (limited) striking surface only. Apart from a minor number of natural 'errors' (flaws) in the flint, also technological errors (step-fractures and hinged terminations) appear in the Area 240 core reduction (*cf.* Shelley 1990).

Characterisation of the Hand Axe Assemblage

Area 240 yielded 33 hand axes of which 31 underwent analysis; two were not available (Appendix 1). Similar to the proportion of the represented cores, the hand axes present are mostly approximately 20 mm longer than wide (Fig. 2.2) and their morphology appears rather homogeneous. The majority (19) of these bifacially flaked pieces retained some cortex, which was less than a quarter of the surface area in each case. Pre-flaking frost fissures are present on only eight hand axes. Overall, the tip of the bifaces is slightly rounded in 11 cases, while a trimmed rounded base is predominant, found in 18 hand axes. A symmetrical 'hand axe shape' prevails on 18 of the

Table 2.1 Quantitative data on the flake tool typology

Type	No.	Artefact
Single convex scraper, type 10 (Bordes 1961)	1	240/18-03-08/060
Side scraper	2	240/07-12-07/006; 240/11-02-08/056
Steep scraper/denticulate ?	1	240/31-01-08/047
Notch (?), type 42 (Bordes 1961)	2	240/17-12-07/005; 240/03-03-08/069
End notch, type 54 (Bordes 1961)	1	240/17-01-08/016
Retouched piece (?), type 99 (De Loecker and Schlanger 2006)	4	240/17-12-07/004; 240/04-02-08/048; 240/24-01-08/026; 240/15-02-08/071
Pieces with macroscopic signs of use (?), type 98 (De Loecker and Schlanger 2006)	2	240/16-01-08/019; 240/07-02-08/058
Total	**13**	

Table 2.2 Typological summary of the hand axes

Typology (after Bordes 1961, where applicable)	No.	Artefact
Subtriangular hand axe	2	240/07-12-07/002; 240/17-01-08/020
Cordiform hand axe	8	240/17-12-07/001; 240/04-01-08/007;240/17-01-08/015; 240/21-01-08/030; 240/31-01-08/076;240/17-12-07/079; 240/18-03-08/089; 240/07-02-08/090
Between cordiform and ovate hand axe	1	240/11-01-08/010
Sub-cordiform hand axe	6	240/22-01-08/027; 240/22-01-08/028; 240/24-01-08/036; 240/13-02-08/052; 240/14-01-08/078; 240/29-01-08/080
Elongated sub-cordiform hand axe	1	240/25-01-08/037
Discoide à talon hand axe	1	240/14-03-08/085
Between *amygdaloid à talon* and *ovalaire à talon* hand axe	2	240/07-02-08/057; 240/22-01-08/077
Broken, probably cordiform hand axe	2	240/21-01-08/029; 240/21-01-08/031
Broken, probably cordiform or sub-cordiform hand axe	1	240/13-02-08/051
Broken hand axe (flat) – tip	2	240/17-12-07/003; 240/17-01-08/023
Broken hand axe (flat) – base	1	240/24-01-08/024
Broken hand axe (flat)	4	240/11-01-08/011; 240/04-02-08/044; 240/12-02-08/053; 240/15-02-08/059
Total	31	

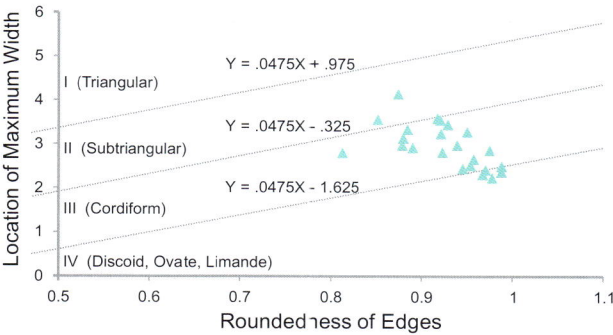

Figure 2.3 Graphical representation of the hand axe types (location of the maximum width versus the roundness of the edges) according to Bordes (1961). The scatter diagram shows four zones which are separated by the dotted lines; an arbitrary division proposed by Bordes (1961) to distinguish four classes of flat bifaces (triangular, sub-triangular, cordiform and discoid/ovate/limande). Zone IV is generally considered as a diverse group. Drawing after Debénath and Dibble (1994)

Based on the hand axe measurements and associated ratios according to Bordes' (1961; Debénath and Dibble 1994:132) the hand axes were typologically classified (Table 2.2, Figs 2.3 and 2.4; see also De Loecker 2010). The Area 240 hand axe assemblage is remarkably homogeneous with cordiform (Pl. 2.4, A) and sub-cordiform (Pl. 2.4, B) hand axes prevailing. In general, the Area 240 hand axe assemblage can be described as Acheulean (Goren-Inbar and Sharon 2006) or as 'Mousterian of Acheulean Tradition' (MTA) tradition, but without backed knives (Bordes 1961).

Raw Material Procurement

The majority of artefacts indicate that rolled flint nodules must have been collected by early hominins from river deposits, such as terraces or exposed gravel bars. Artefacts made from these raw materials show abraded cortex and 'old', patinated and rounded, natural fractures. Yet, a small number of artefacts are characterised by a more 'fresh-looking' chalky cortex, or a rough cortex with cavities. The latter class of artefacts, together with large flake dimensions (which imply large raw material nodules), could suggest that primary flint sources were available. It appears that the raw materials had already been affected by frost before knapping took place. A significant number of the assemblage, 30 artefacts, display 'fresh' natural flaws indicating either a non-selective choice of raw material or a lack of 'high' quality raw material, together with the absence of testing. However, the raw materials used for the production of the hand axes are, in general, of higher quality than the frost affected flint.

artefacts, where both working faces are mainly convex. The group of tools largely shows a continuous pattern of retouch on 17 artefacts. Five bifaces (240/17-12-07/001; 240/07-12-07/002; 240/13-02-08/052; 240/14-01-08/078 and 240/17-12-07/079) show a tip which was shaped by a tranchet blow (the precise removal of a large flat flake from the tip of the biface). On the whole, these tranchet removals could suggest that the given hand axes were sharpened or resharpened. Soft-hammer percussion using softer material such as wood, bone or antler, was predominately used for hand axe trimming.

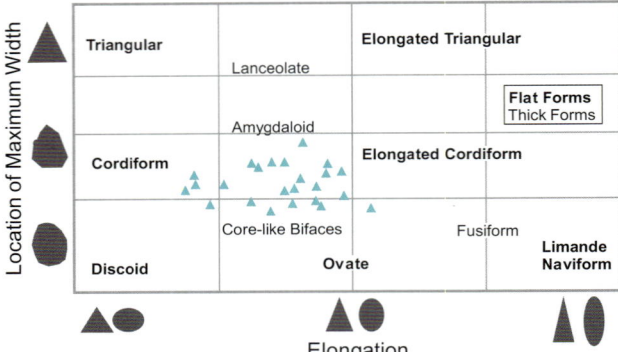

Figure 2.4 Graphical representation of the hand axe types (location of the maximum width versus the elongation) according to Bordes (1961). Drawing after Debénath and Dibble (1994)

On the basis of texture, cortex, inclusions and colour, the flint can be characterised roughly as a dark-grey to light-grey flint, with characteristic inclusions in the form of (small) lighter coloured dots, spots and stains. The flint is typified as fine-grained, but also shows larger 'circular' (oval) coarser-grained zones, inclusions and fossils. At the edges of the nodules (underneath the cortex), the flint shows a darker-grey/grey-bluish zone, which is more transparent (glassy) and often shows small amounts of bluish-white patina. Generally, the used flint is homogenous in character, with major variations in colour. While it is difficult, if not impossible, to relate the flint varieties to specific, pinpointed geographical sources, based on the homogeneity of the flint, it is likely that the source material were present locally, in nearby gravel beds, rather than transported into the area.

Post-depositional Artefact Modifications

The condition of the artefacts indicates that several taphonomic processes were responsible for the formation of the assemblage. Generally, a 'heavy' patina, with variations in colour and gloss, was present on the artefacts. Forty-nine artefacts show small collision and pressure cones, which are few in number and are mainly observed near ridges between flake scars. Twenty-five of the artefacts exhibit some natural scratches on flatter parts. The assemblage also shows some dulling of the edges, while only 13 of the artefacts show heavy rolling all over, including flake scar ridges. The same number of artefacts shows a network of incipient frost fissures (lines) which cut the flake scars, while eight have a partly missing surface due to frost cracking after flaking. Pre-manufacture flaws are evident on only eight of the hand axes. All artefacts exhibit some sign of fresh damage; commercial dredging and processing of the gravels is the likely cause. In all, the data indicate that there are obvious differences in abrasion and preservation within the assemblage. Moreover, part of the assemblage seems to have spent a considerable period of time in iron-rich gravels, while a minority of the artefacts show edge and surface conditions which are indicative of burial in fine-grained sediments. The condition and quality of the flint artefacts show that the Area 240 assemblage probably originates from several different contexts:

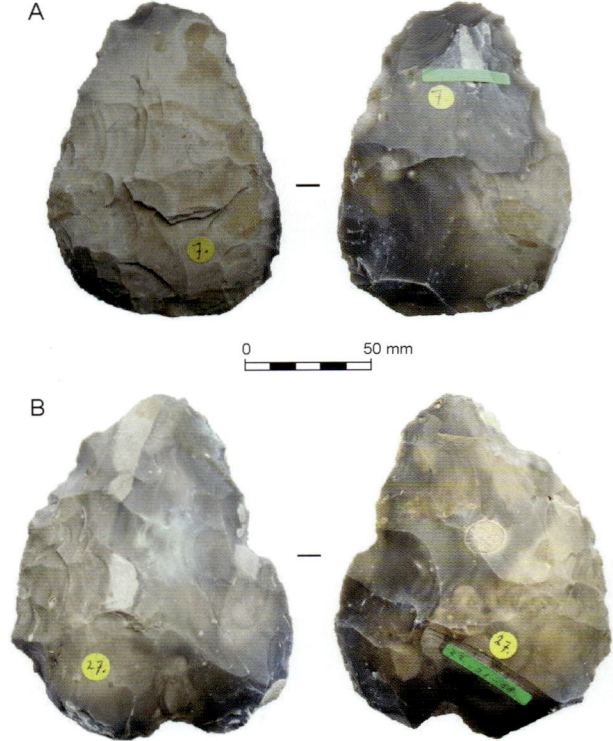

Plate 2.4 A) Example of a cordiform hand axe. B) Example of a sub-cordiform hand axe. Scale 1:2. Photographs by Jan Meulmeester

1. Artefacts which show a (minor) colour patina, a light gloss and insignificant edge damage (among these are some flakes and cores, but mainly the hand axes): since the hand axes show a rather uniform morphology, a (near) primary context situation cannot be excluded.
2. Artefacts which show predominantly post-depositional surface modifications (colour patina, gloss and abrasion) on one single side of the artefact (this group mainly consists of flakes): an eroding surface is suggested.
3. Heavily weathered artefacts (heavy colour patina and gloss, and abrasion and battering on both sides of the artefacts), which probably originate from a secondary context such as gravel deposits.

Nevertheless, it is likely that at least some of the material was dredged from primary contexts. Generally, sediments accumulated during many phases of deposition are suggested. If the artefacts came from a single, discrete location, it is probable that they derive from a fluvial sequence with both finer grained and coarse grained layers such as a gravel body with intercalated fine sandy layers. This hypothesis was tested through the geophysical and sampling phases of the project.

The Area 240 Faunal Remains: a Summary

Faunal remains amounting to approximately 130 fragments were recovered along with the flint artefacts. The assessment was carried out by Jan Glimmerveen and the following results were provided to the authors. The recovered faunal remains include a large number of terrestrial mammal bones: woolly mammoth, woolly rhinoceros, bison, reindeer and horse. The condition of the faunal remains is similar to the flint artefacts in that some are well preserved indicating a possible *in situ* sedimentary context and some abraded, indicating a secondary context prior to recovery.

Approximately 70% of the faunal remains were attributed to an age between 42,000 and 32,000 BP (MIS 3) based on the radiocarbon dating of five bones (Table 2.3). The radiocarbon dating was carried out by Professor Hans van der Plicht, Centre for Isotope Research at Groningen University, Netherlands.

Table 2.3 Radiocarbon dates of faunal remains recovered from Area 240 as part of the original discovery (courtesy of Jan Glimmerveen)

Lab number	Material	Radiocarbon ages (uncalibrated BP)
GrA-39965	Woolly rhinoceros (*Coelodonta antiquitatis*) mandible fragment	>45,000 BP
GrA-39962	Woolly mammoth (*Mammuthus primigenius*) cervical vertebra	37,240 (+280, -260) BP
GrA-39966	Reindeer (*Rangifer tarandus*) antler	31,460 (+160, -150) BP
GrA-39964	Horse (*Equus*) metacarpal	42,960 (+500, -420) BP
GrA-39518	Steppe bison (*Bison priscus*) metacarpal	39,900 (+850, -650) BP

The radiocarbon dating indicates a MIS 3 age and the remains dated comprise the larger mammals indicated by the British MIS 3 Pin Hole mammal assemblage (Current and Jacobi 2001). However, there is also the possibility that the remains could be older. Hijma *et al.* (2012) concluded that radiocarbon measurements of North Sea material cannot distinguish MIS 3 from MIS 4–5 bones in standard dating practice and undervalue their true dates, possibly due to age overprinting from younger organic material that contaminates the dating method. Briant and Bateman (2009) came to similar conclusions when comparing radiocarbon with OSL dates from Late Pleistocene fluvial sequences in eastern England. That study found that OSL and radiocarbon dates agreed well for ages younger than *c.* 29 ka ^{14}C BP (*c.* 33 ka cal BC) but disagree beyond *c.* 35 ka ^{14}C BP (*c.* 38 ka cal BC) which they suggested was due to radiocarbon ages on older organic material having been contaminated by low levels of modern carbon. It is therefore suggested that all conventionally pre-treated radiocarbon ages greater than *c.* 35 ka ^{14}C BP should be considered with caution. However, given the cool temperature faunal remains that were selected for dating, it is likely that they are of early to mid-Devensian age.

The remaining 30% of the bones are heavily fossilised, estimated to be older than 500 ka (J. Glimmerveen pers. comm.). However, preservation of bones, and the degree of fossilisation may vary within a deposit and between deposits. As it is unknown if the bones were from a single sediment deposit or from many deposits, these dates should not be taken as definitive.

Discussion

The results of the lithic analysis and faunal assessment suggest a palimpsest situation, created by cultural as well as natural site formation processes (eg, Schiffer 1976; 1983; 1987; Shipman 1981; Hodder 1986; Nickens 1991):

1. The assemblage was dredged from a provenance area which covers approximately 3.5 x 1 km.
2. The finds were recovered amongst river-terrace sands and gravels, on an outsize pile at an onshore commercial sorting centre.
3. The lithic collection is formed by three groups of artefact which show a specific combination of secondary surface modifications, such as patina, gloss, rolling, frost actions, etc. (*cf.* Stapert 1976).
4. The assemblage represents a morphological mixture of artefacts which were fashioned

during several different technological manufacturing processes (eg, hand axe production, prepared Levallois technology and disc/discoidal core approach).

Positive proof for this assumption is, amongst others, given by the fact that none of the recovered flakes can directly be related to hand axe production or (re)sharpening. Moreover, the artefacts correspond to several different stages of *chaînes opératoires* (eg, flakes, flake tools, exhausted cores, hand axe-like rough-outs, ´finished´ hand axes, etc.). A large diversity in the products and waste of the core reduction is noticed. Crude and thick flakes, as well as, large well-prepared Levallois flakes and cores are jointly recovered. All in all, they are generally struck from large flint nodules. On the whole, initial stages of core reduction (decortication) are lacking, whereas the original *chaînes opératoires* are only represented by fragments. The assemblage shows that rather large and wide, but 'thin', nuclei were discarded. Compared with the large flake measurements, it can be suggested that the cores were not directly responsible for the production of the flakes. Although in an alternative scenario, the flake assemblage could represent former steps in the reduction, while the final stages of flake production (smaller flakes) were lost. Metric data and morphology shows that the recovered flakes could not have resulted from the final stages of hand axe preparation. If there is a correlation between the flakes and the hand axes, it will be in the initial phases of biface production. Moreover, large flakes are frequently used as blanks to produce hand axes (or rough-outs). Conspicuously, however, the majority of lithics show a rather homogeneous type of flint and the hand axe assemblage looks uniform.

The Area 240 lithic assemblage can be considered typologically heterogeneous. However, it is not possible to estimate the age of the artefacts based on typological classification alone. Typological series must always be treated with caution, as certain artefact types are found commonly throughout the Palaeolithic (*cf.* Woodcock 1978). The finds are characterised by the occurrence of cordiform or sub-cordiform hand axes, and include a series of well-made Levallois (*sensu stricto*) products. Together with the post-depositional conditions of the artefacts, the mixed assemblage is composed of different Acheulean and/or Mousterian elements.

Frequently, the Acheulean is characterised by hand axes and cleavers (Goren-Inbar and Sharon 2006), while a *façonnage* style of reduction is suggested. This implies that the focus of lithic reduction was not on the production of flakes and debris (*débitage*), but on transforming a raw material nodule into a tool (hand axe: bifacial façonnage is organized to create two convex surfaces). The earliest secure evidence shows that the British mainland was already occupied before 780 ka. Yet, the lower Palaeolithic assemblages at Pakefield (MIS 17, *c.* 710 ka) and Happisburgh constitute a straight-forward core and flake industry (Parfitt *et al.* 2005; 2010) and are therefore not representative of classic Acheulean material. According to a number of well-dated key sites (eg, Boxgrove, West Sussex (Roberts and Parfitt 1999); High Lodge, Suffolk (Ashton 1992); Swanscombe, Kent (Wenban-Smith and Bridgland 2001); Hoxne, Suffolk (Ashton *et al.* 2008); Red Barns, Portchester (Wenban-Smith *et al.* 2000); and Elveden, Suffolk (Ashton *et al.* 2005)) large hand axe dominated assemblages appear in the British archaeological record somewhere around MIS 13–12 (*c.* 533–424 ka), and continue to appear up to MIS 9–8 (*c.* 337–243 ka). These typical Acheulean industries are, above all, dominated by a wide range of ovate, pointed and cordate hand axe assemblages. Nevertheless, non-hand axe flake-tool and -core industries occur as well throughout this period (*cf.* Roe 1967; 1968; Wymer 1968; Ashton 1992; Ashton *et al.* 1992; McNabb 1992; Ashton and McNabb 1994; White 1998a–b; Roberts and Parfitt 1999; Wenban-Smith *et al.* 2000; Wenban-Smith and Bridgland 2001; Wenban-Smith 2004; Ashton *et al.* 2005; Ashton *et al.* 2008; Emery 2010). It is therefore possible, although very speculative and only based on a limited comparison, that the Area 240 hand axes represent the remnants of a Lower Palaeolithic Acheulean findspot (roughly between *c.* 500 and 300 ka), though there are no arguments, based on typology alone, to exclude a younger date.

The Lower–Middle Palaeolithic transition is conventionally marked by the emergence of Levallois-dominated core technologies, which appear from around 300 ka (late MIS 9–8) onwards (*cf.* White and Ashton 2003). Subsequently, flake-orientated Levallois products became more and more predominant from MIS 8 and 7 (*c.* 243–191 ka) (White *et al.* 2006; Emery 2010; Scott and Ashton 2011). This could imply that the Area 240 Levallois products and the well-made hand axes are broadly contemporaneous in geological terms: a statement supported by the homogeneous character of the used flint as well as the post-depositional surface condition on some of the artefacts. In this scenario the dredged artefacts (or at least the Levallois component) date in the region of 300 ka or younger. However, resolving the artefacts to an archaeological context (ie, a defined layer of sediment) is not possible at Area 240 where we must be content with attributing material to a geological unit (ie, multiple layers of sediment and potentially multiple archaeological contexts), as discussed in later chapters. Like Area 240, a number of British sites (eg, Ebbsfleet, Kent (Wenban-Smith *et al.* 2006); Lion Tramway Cutting at West Thurrock,

Essex (Schreve *et al.* 2006); Pontnewydd Cave, Clwyd (Wales) (Pettit and White 2012); Harnham, Wiltshire and Broom, Devon (Ashton and Hosfield 2009)) contain both hand axe and Levallois industries in taphonomically complex sedimentary deposits.

After early MIS 7 (*c.* 243 ka) a scarceness of British sites is observed in the record (*cf.* Ashton 2002; Ashton and Lewis 2002; Scott and Ashton 2011). During the last glacial cycle (MIS 5d to 3; *c.* 115 and 29 ka), hand axes appear again in the so-called 'Mousterian of Acheulean Tradition' (MTA). Principally the MTA shows thin, symmetric (sub-) cordiform or (sub-) triangular hand axes, while in Britain *bout coupé* hand axes are found repeatedly (White and Jacobi 2002; Boismier *et al.* 2003). On average, these Mousterian bifacial elements are generally smaller than the Lower Palaeolithic examples (Bordes 1961; Cliquet and Lautridou 1988; Soressi 2002). Both cordiform and sub-cordiform hand axe types could belong to MTA assemblages, the measurements of the Area 240 bifaces are rather large, although large hand axes have been observed within the Lynford Quarry assemblage (Boismier *et al.* 2012). The presence of several Levallois products indicate that the assemblage is, at least in part, Early Middle Palaeolithic. The hand axes are thought to be Acheulean, however a younger MTA interpretation cannot be ruled out.

In addition, the relationship between artefacts and faunal remains is difficult to define. The faunal assessment indicates two groups relating to the Early and Late Pleistocene. Although, some of the faunal remains are in good condition and were possibly *in situ* prior to recovery, other faunal remains are broken, abraded and are clearly from secondary contexts (Jan Glimmerveen pers. comm.). It is considered likely that there is no direct relationship (apart from being recovered together) and is a result of post-depositional processes, both natural as well as industrial. Furthermore, subsequent recovery of terrestrial faunal remains from Area 240 (discussed further in Chapter 6) indicate a wider age range of faunal material with late Middle Pleistocene and early Holocene remains recovered in addition to Early and Late Pleistocene remains. This further attests to the palimpsest situation.

In conclusion, the Area 240 assemblage recovered from the wharf in 2008/2009 can be considered as a mixed group of artefacts, or better defined as a palimpsest of mixed ages, representing a variety of technological actions, and can be described as Levallois and, Acheulean and/or as MTA. However, the tear-drop symmetric-shaped hand axes are particularly uniform both in terms of technology and typology, and the nature and quality of the used flint seems to be remarkably homogenous for all artefacts. A palimpsest situation, created by cultural and natural site formation is concluded; the complexity of which is not necessarily surprising, as examples of similar assemblages have been noted within British contexts and as a result of the mode of collection, sorting and ultimate discovery. Similarly, the initial assessment and dating of the faunal remains indicates a range of ages indicating the intermittent presence of terrestrial mammals ranging in date from Cromerian (>500 ka) to late Devensian (*c.* 30 ka), deposited in a range of sediments.

The recovered assemblage of flint material and faunal remains should be treated as a complex 'surface scatter' with a restricted value for early human behavioural inferences (*cf.* Wymer 1995; Kolen *et al.* 1999), based on the material alone. The further defining and understanding of the depositional (contextual) regime and the systematic controlling of the recovery of further finds in Area 240 was therefore of crucial importance for the interpretation of the assemblage. The following chapters detail the methodology and results of the work carried out in order to assess the context of the assemblage, and further recovery of artefacts and faunal remains from within Area 240. Many of the themes outlined in this chapter with regards to taphonomy and resource material are returned to in Chapter 6 where the entire collection of material from Area 240 is discussed in its geological context.

Chapter 3
Palaeogeographic Reconstruction Methods

Introduction

Prior to this project relatively little was known about the geology of Area 240 and the surrounding aggregate block; it was not possible to describe the context for the recovered archaeological material in any detail. Previous work in the wider offshore region by the British Geological Survey (BGS), based on geophysical datasets acquired during the 1970s and 1980s, indicated that the area is dominated by Early Pleistocene shallow marine sands partially overlain by early Devensian channel deposits with the seabed dominated by marine sands and gravels (British Geological Survey 1991). However, Bellamy (1998) intimated that the sub-surface sand and gravel deposits were associated with a now-submerged downstream section of the River Yare and that the sediments were equivalent to the onshore Yare Valley and Breydon Formations. A later study carried out to the north and adjacent to Area 240 (Wessex Archaeology 2008c) further indicated the complexity of deposition and erosion in the area, with potential Saalian and early Devensian deposits present. Previous work commissioned by the licensee within Area 240 itself was carried out with the aim of assessing the aggregate resource (presence and thickness) rather than the development of deposition, erosion and environments. There were obvious gaps in knowledge of the palaeogeography of Area 240 and more generally within the East Coast aggregate licence block. It was clear that a reconstruction of the landscape was required in order to ascertain the context of the archaeological material recovered. A focus was placed on the age and type of depositional environments and subsequent erosional history caused both through natural and man-made processes.

In *Submerged Forests*, Reid (1913) produced the earliest reconstruction of the North Sea landscape at lowered sea levels based on evidence of coastal peats and submerged forest. This early attempt to answer questions on past sea level, climate and vegetation at the end of the Last Glacial period resulted in the production of a map which illustrated the contemporary coastline with the lowest of the known submerged forests. In what would become a first attempt at mapping the Mesolithic landscape, Reid illustrated the extensions of European river systems across the 'alluvial flat' that connected Britain and Europe.

Subsequently, Clark (1936) provided a synthesis of the Mesolithic archaeological and environmental evidence from Britain, Europe and the North Sea Basin. The development of the submerged landscape was augmented by pollen data recovered from submerged peat and a harpoon point within the vicinity of the peat (Godwin and Godwin 1933). Clark's subsequent work into the 1970s continued to treat the North Sea as a landscape although subsequent discussions during the 1970s and 1980s tended to treat the North Sea Plain as a land-bridge and as an unfavourable environment; a concept which continued until the late 1990s (for further discussion see Coles 1998).

Coles (1998) detailed the development of these concepts from Reid's work onwards describing the change in opinion from thinking of the North Sea plain as a land-bridge to envisioning an expansive plain that was available for use and settlement. Coles (1998) produced a series of maps providing speculative reconstruction of the topography, river systems, shoreline configurations and occupation from the Last Glacial Maximum to the start of the Neolithic. These reconstructions were essentially based on limited, low resolution data including bathymetry and general contour maps of the seabed. Although large topographical features that affect the shape of the seabed were considered, similar scale, albeit buried features (that have since been shown to exist) were not identified, nor discussed in any detail. Historically, such broad scale reconstructions (eg, Coles 1998; Fleming 2002; Wenban-Smith 2002) have provided high quality syntheses of specific archaeological periods and for certain regions. However, they tend to over-simplify the geological modification processes operating on the longer-term archaeological record, specifically in terms of multi-period, syn- and post-transgressive site formation (for discussion see Westley *et al.* 2004). Over the past decade data availability has increased both in quality and quantity and, as such, has vastly increased our ability to reconstruct landscapes and to assess their potential use by humans throughout periods of known occupation.

Recent exploration of the UK Continental Shelf has identified vast submerged landscapes dated to various periods within the timescale of human occupation. These landscapes have been reconstructed at a variety of scales from large regional studies such as the Regional Environmental Characterisation (REC) surveys (EMU Ltd 2009; Limpenny et al. 2011; Tappin et al. 2011; James et al. 2010) and the *North Sea Palaeolandscapes Project* (Gaffney et al. 2007) to smaller-scale, high resolution studies (Wessex Archaeology 2008a–f; Mellett et al. 2012). These studies have used geophysical and geotechnical datasets as their primary sources. The Mesolithic material from Bouldnor Cliff in the Solent, off the Isle of Wight (Momber et al. 2011) has allowed for the study of a prehistoric landscape at the site-level as well as the development of underwater archaeological techniques in challenging working conditions.

The methodologies and implementations of these methodologies vary and as such the resolution of reconstruction also varies. Small-scale localized studies can provide a level of detail not necessarily attainable on larger, more regional studies. However, a high level of detail in a small area is of little use without a more regional palaeogeographical and archaeological context.

Regardless of scale or resolution, most reconstructions in UK waters have been used primarily to assess the potential for archaeology within their respective regions. The resulting assessment of potential is largely based on the age and specific features of a landscape, and knowledge of cultural attributes relating to human occupation and land-use during specific periods. Modelling can be based on terrestrial analogues, ethnographic examples or technologically contemporary cultures found elsewhere in the archaeological record.

Area 240 Investigations

The discovery of archaeological material from an isolated zone within Area 240 allowed for the development of a project that aims to reconstruct a landscape based on actual, recovered archaeological material rather than speculative, theoretical potential. Existing geophysical and geotechnical methodologies were used to acquire data with sampling strategies adjusted to suit the nature of the project. The goal was to refine practical techniques used to establish the presence of archaeological, palaeontological and palaeoenvironmental material, and to establish character, age and extents of sediment units associated with the discovery of flint and faunal remains.

The project was developed in a series of stages allowing the work to be evidence-based, with each stage informed by the results of the preceding stage. A major advantage of an evidence-based, staged approach was the continued development of subsequent stages of the project rather than having been designed around a series of assumptions formed at the project's inception. This area of the southern North Sea has strong seabed currents, predominantly oriented north–south, with renowned poor visibility, and environmental conditions were considered throughout the programme. These conditions dictated, in part, what methodologies were chosen and also how the methods were implemented. The design and development of the project implemented the appropriate lessons learned during the project *Seabed Prehistory: Gauging the effects of Marine Aggregate Dredging* project (Wessex Archaeology 2008a–f). Details of the investigations carried out in Area 240 are specified below.

Stage 1: Existing data review

The *Existing Data Review* comprised the processing and interpretation of an existing geophysical dataset and the integration of geotechnical data within the area (Wessex Archaeology 2009a). The aim of this stage was to provide geological context for Area 240 with specific regard to the sediment units from which the artefacts were most likely to have been dredged. This was accomplished by the assessment and integration of geophysical and geotechnical data provided by Hanson Marine Aggregate Limited. The geological assessment focused on those sediments thought to have been deposited within the timescale of the human occupation of Britain (ie, approximately the last one million years).

The geophysical dataset was originally acquired in 2005 for the purposes of aggregate resource assessment. The data were acquired by Andrews Survey Limited (now Gardline Marine Surveys) and comprised sub-bottom profiler (boomer source) and multibeam echosounder (swath bathymetry) data. All data were acquired on north–south orientated lines at 100 m line spacing, with cross-lines acquired at 1 km line spacing (Fig. 3.1).

The sub-bottom profiler data were acquired using an EG&G 230 surface-tow boomer with external hydrophone. Positioning was provided by DGPS with a fixed layback of 30 m. Data were of very good quality, particularly the north–south orientated lines. The east–west orientated lines are of slightly poorer quality, likely caused by the strong cross-currents in the area having an adverse effect on the data. The data processing and interpretation was carried out using Coda GeoSurvey software. The sub-bottom profiler data were interpreted with two-way travel time (TWTT) along the z-axis. In order to convert from TWTT to depth, the velocity of the seismic waves was estimated to be 1,600 m/s, a standard estimate for shallow, unconsolidated sediments (Sheriff and Geldart 1983). This value was consistently used throughout subsequent interpretations.

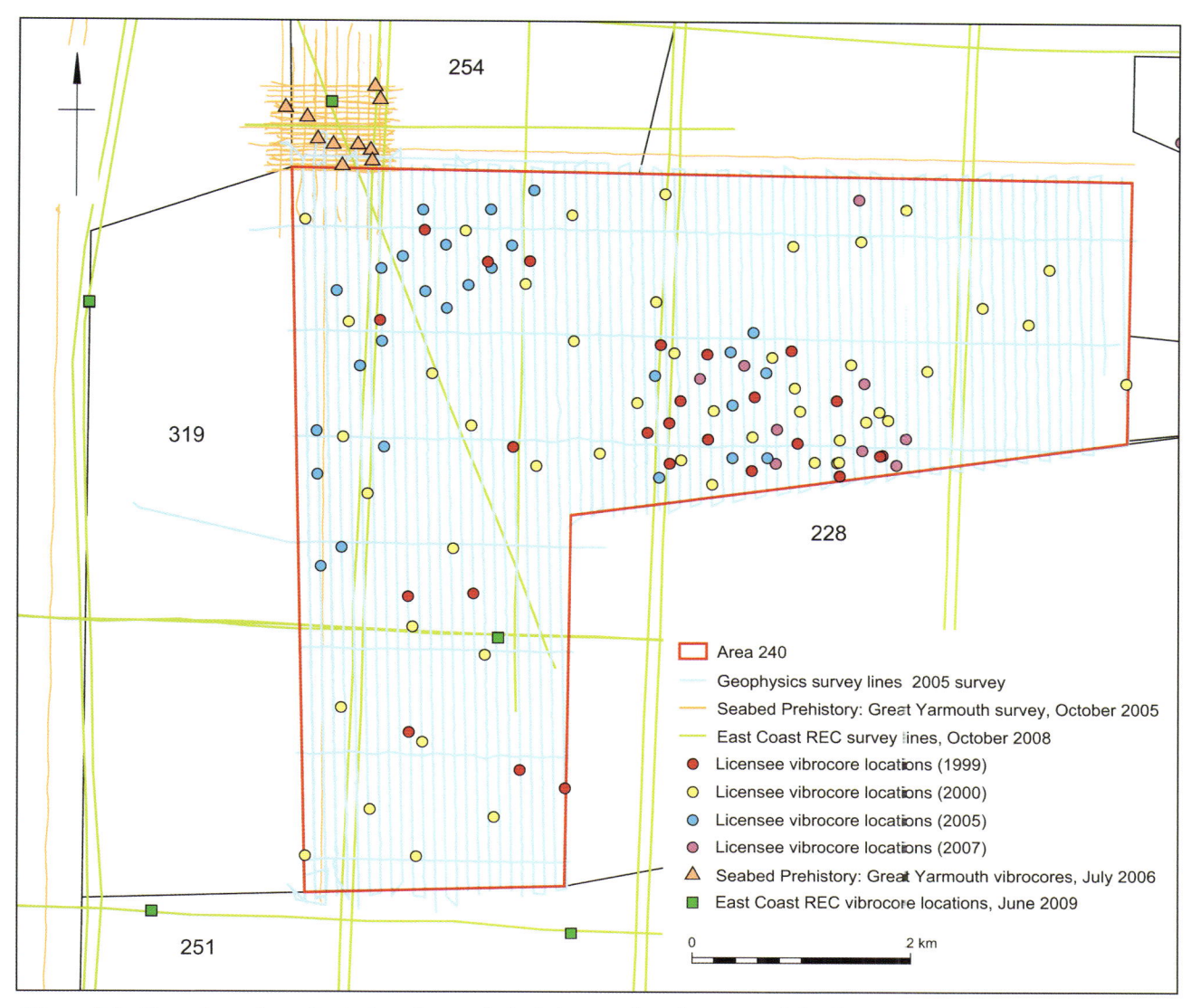

Figure 3.1 Overview of geophysical and geotechnical data reviewed as part of the Stage 1 assessment

The interpretation of sedimentary boundaries and features were exported from the Coda GeoSurvey software and imported into ArcGIS, which was used to integrate the different datasets and map the entire project. The depth of boundaries sub-seabed were also exported and gridded into layers using IVS Fledermaus software, referenced to the seabed depths acquired during the bathymetry survey (reduced to metres below Ordnance Datum (OD)). This allowed 3-D assessment of the interpretation across the area.

The bathymetry data were acquired using a Geoswath system and the data were supplied as a tidally corrected gridded SD file (the file format generated by IVS Fledermaus software). Software and processing settings used in this stage were used throughout the project.

As part of the geotechnical data review 109 vibrocore logs (Fig. 3.1) and photographs were reviewed from five surveys undertaken in Area 240 (Alluvial Mining Ltd 1999; Andrews Survey 2000a; 2000b; 2005; Lankelma Andrews 2007). These surveys were commissioned by Hanson Aggregate Marine Limited in order to assess the economic viability of extracting sand and gravel from within the first few metres of sediment below the seabed in Area 240.

Interpolation between geotechnical logs acquired by different contractors during numerous surveys acquired in different years present difficulties, particularly in an area where sands and gravels dominate the sedimentary sequence such as Area 240. The logs were completed at different times by different companies and the recording of the information in the logs can differ. For example, the degree of sorting, sedimentary structure, and gravel type are either not recorded or recorded intermittently within the logs, thus increasing the difficulty when comparing data. The sediments incorporated within deposits are likely to have been originally transported by shallow marine, fluvial and glacial processes and sediments from these sources

> **Box 3.1: Sub-bottom profilers**
>
> Sub-bottom profilers are used to image the sub-seabed geology. A seismic source is deployed and triggered at a fixed firing rate in pings per second (Hz). It emits seismic energy which travels through the water column and penetrates the seabed; some of the energy is reflected back to the surface as it encounters layers with different density and velocity. The reflected energy is then detected by transducers and recorded. Seismic sources can be surface towed (eg, boomer and surface-towed pinger), sub-surface towed (eg, chirp) or hull-mounted (eg, hull-mounted pinger and parametric sonar); seismic transceivers can be either deployed separately (eg, hydrophone for boomer) or are combined with the seismic source as a single piece of equipment (eg, pinger, chirp and parametric sonar). The effectiveness of a sub-bottom profiler system depends on several factors: penetration (determined by power and frequency of the source), positional accuracy, and lateral and vertical resolution. Lateral (or horizontal) resolution is the measure of accuracy of separating sub-surface features and is dependent on the number of data readings and the footprint of the seismic source. The footprint of the seismic source is the area ensonified by the seismic signal and depends on the directivity of the seismic source. Vertical resolution is the measure of accuracy to which vertical reflectors can be separated and is mainly determined by the frequency of the source. Typically, the lower the frequency, the lower the vertical resolution.
>
> In general, data quality mainly depends on suitability of the sub-bottom profiler to sediment composition, water depth and weather conditions during acquisition.

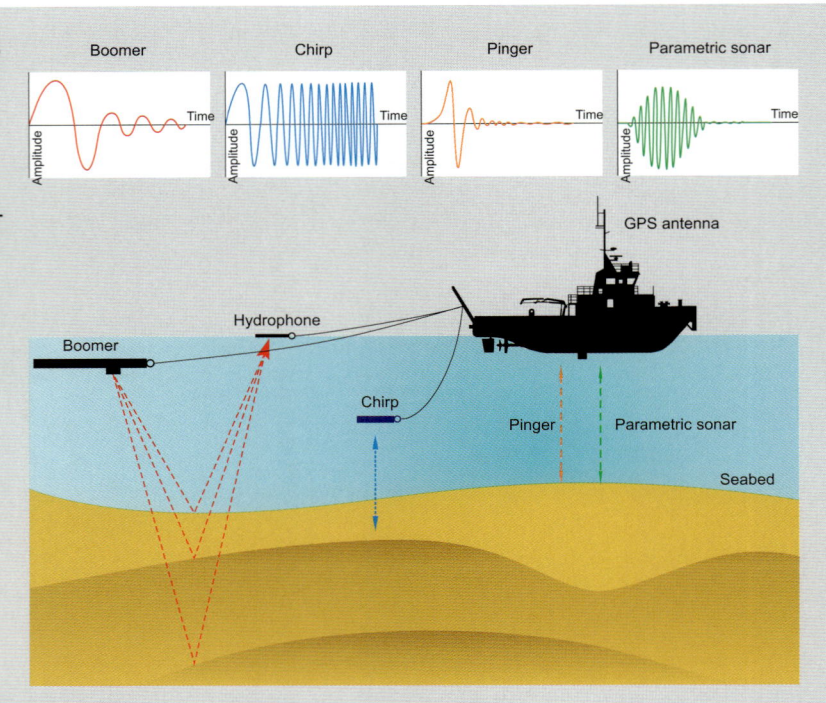

are difficult to differentiate in photographs. Also, the mode of deposition of sand and gravel either by marine, coastal, fluvial and/or glacial processes cannot often be determined by only reviewing geotechnical vibrocore logs. Such specific related information is not always recorded.

Vibrocore data are often provided without water depth information. Normally this is not necessarily a problem if the vibrocores were acquired at the same time as a geophysical survey whereby the bathymetry data can be used as a datum for the top of the vibrocore, as with the 2005 geotechnical dataset. However, the remaining vibrocores were acquired during periods without geophysical data acquisition. Where these vibrocores were acquired without accurate water depth information the 2005 bathymetry dataset has been used as a baseline datum. However, these depths are treated as indicative due to dredging activities and natural movement of mobile sediments within the Area. As the quantity of sediments dredged in intervening years is unknown, fully accurate conversions cannot be made. These issues do not prohibit successful integration of geotechnical and geophysical data; however, care and circumspection is required when referring to depth and thickness of sediment units. Also, dredging activity has likely disturbed the top (approximately 2 m) layer of sediment in some areas and this is not always obvious from the vibrocore logs or photographs. The similarities between the contents of the sediment units meant that when integrating the vibrocore data with the geophysical data, the sedimentary structures were the principal focus of the assessment, supplemented by the sediment record.

In addition to the licensee-provided data, where appropriate, two further datasets were integrated into the interpretation (Fig. 3.1).

In 2005, as part of the *Seabed Prehistory* project, Titan Environmental Surveys Limited were commissioned to undertake a geophysical survey in the East Coast aggregate block (Wessex Archaeology 2008c). Between 30 August and 19 October 2005, sub-bottom profiler (surface-tow boomer) data were acquired along five regional prospection lines (65 line km). This was followed by an 800 x 800 m survey in the south-west corner of Area 254 situated adjacent to Area 240 (Fig. 1.3). Sub-bottom profiler (surface-tow boomer and pinger), sidescan sonar and single-beam echosounder data were acquired along east–west orientated lines at 50 m spacing.

The use of two sub-bottom profiler sources (boomer and pinger) was dictated by the anticipated sediments in the area. Bellamy (1998) indicated that the area was dominated by sands and gravels overlain by a fine-grained unit, comprising peat and other organic matter. Coarse-grained sediments are better resolved using the boomer source, whereas the pinger source is better at determining fine-grained sediments. This use of multiple sources was further investigated in Stage 2 of the Area 240 project.

The use of 50 m line spacing was implemented following the results of the earlier study in the Arun Palaeovalley as part of the *Seabed Prehistory* project (Wessex Archaeology 2008a). The project demonstrated the need for narrow line spacing in order to map small features. After comparing the interpretation at a variety of grid sizes (200 x 200 m, 100 x 100 m and 50 x 50 m) it was concluded that surveys undertaken with a line spacing of no more than 100 m with cross lines situated up to twice the principal line spacing provided a quality dataset for interpretation (Fig. 3.2). However, a reduced line spacing of 50 m for the principal orientation significantly improved the resolution and therefore clarity of the interpretation with smaller features identified (Wessex Archaeology 2008a).

Due to the close proximity of the survey area to Area 240 the original interpretation was re-visited in parallel to the Area 240 assessment in order to provide a coherent interpretation for both areas. A vibrocoring programme in Area 254, based on the results of the geophysical data, was conducted by Gardline Surveys Limited in 2005 on behalf of Wessex Archaeology. The cores allowed for the calibration of the identified geophysical horizons and provided samples for environmental analysis and dating. A total of 16 vibrocores were acquired at eight locations; one core at each location was acquired for Optical Stimulated Luminescence (OSL) dating. Samples from two vibrocores (Vc1 and VC3) underwent palaeoenvironmental assessment, analysis and dating (Wessex Archaeology 2008c).

The second relevant dataset was acquired during the East Coast REC project, which was a multidisciplinary investigation that employed a variety of techniques to develop a broad understanding of the habitats and areas of archaeological interest over an extensive area of approximately 3300 km² of the seabed off the East Anglian coast (Limpenny *et al.* 2011). The study was conducted over a three-year period and was funded by the Marine Aggregate Levy Sustainability Fund. The geological and archaeological outcomes of the project provided regional context to the palaeogeographic reconstruction of Area 240.

Box 3.2: Swath bathymetry

Multibeam echosounder (or swath bathymetry) data measures water depth with a fan-shaped array of acoustic beams that extend below and to the sides of the transducer, usually attached to the hull of the survey vessel, to acquire a swath of spot depths. The swath width is typically around four times the water depth. As the vessel moves forward continuous and well-positioned spot depths are recorded allowing a digital terrain model to be produced. Natural seabed features, such as topographical changes, bedforms etc. and man-made features such as the dredging scour and wreck material can be observed and spatially mapped.

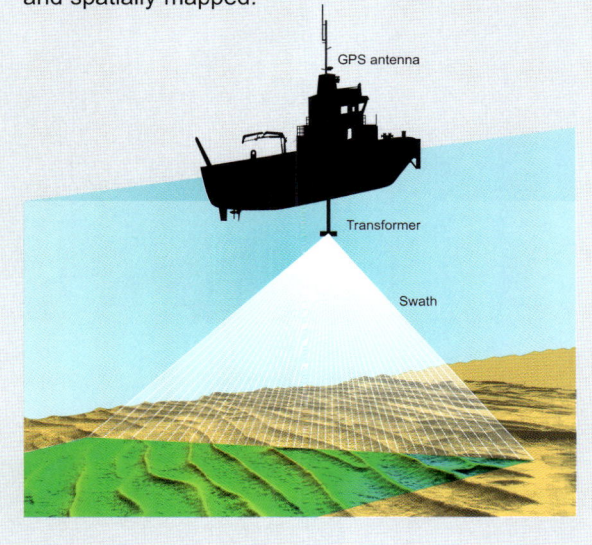

The REC geophysical survey was conducted on board the RV *Cefas Endeavour* during September and October 2008. It comprised 20 north–south corridors (each comprising two survey lines) situated approximately 2.7 km apart and 26 east–west oriented lines. This produced a survey grid of approximately 3315 line kilometres and provided coverage for the 3300 km² REC area. Sidescan sonar, sub-bottom profiler (boomer), multibeam echosounder, magnetometer and acoustic ground-discrimination system data were acquired. In addition to the corridor survey, a further three survey lines locations were designed to specifically support the work carried out in parallel at Area 240 and were designed, based on the current thinking at that time, to focus on the palaeogeography of the Area.

During May and June 2009 the RV *Cefas Endeavour* returned to the REC survey area and conducted a ground-truthing survey with the overarching aim of validating the geophysical interpretation data. The survey comprised the acquisition of geotechnical samples (vibrocores and clamshell grabs) and biological sampling (benthic grabs, camera images,

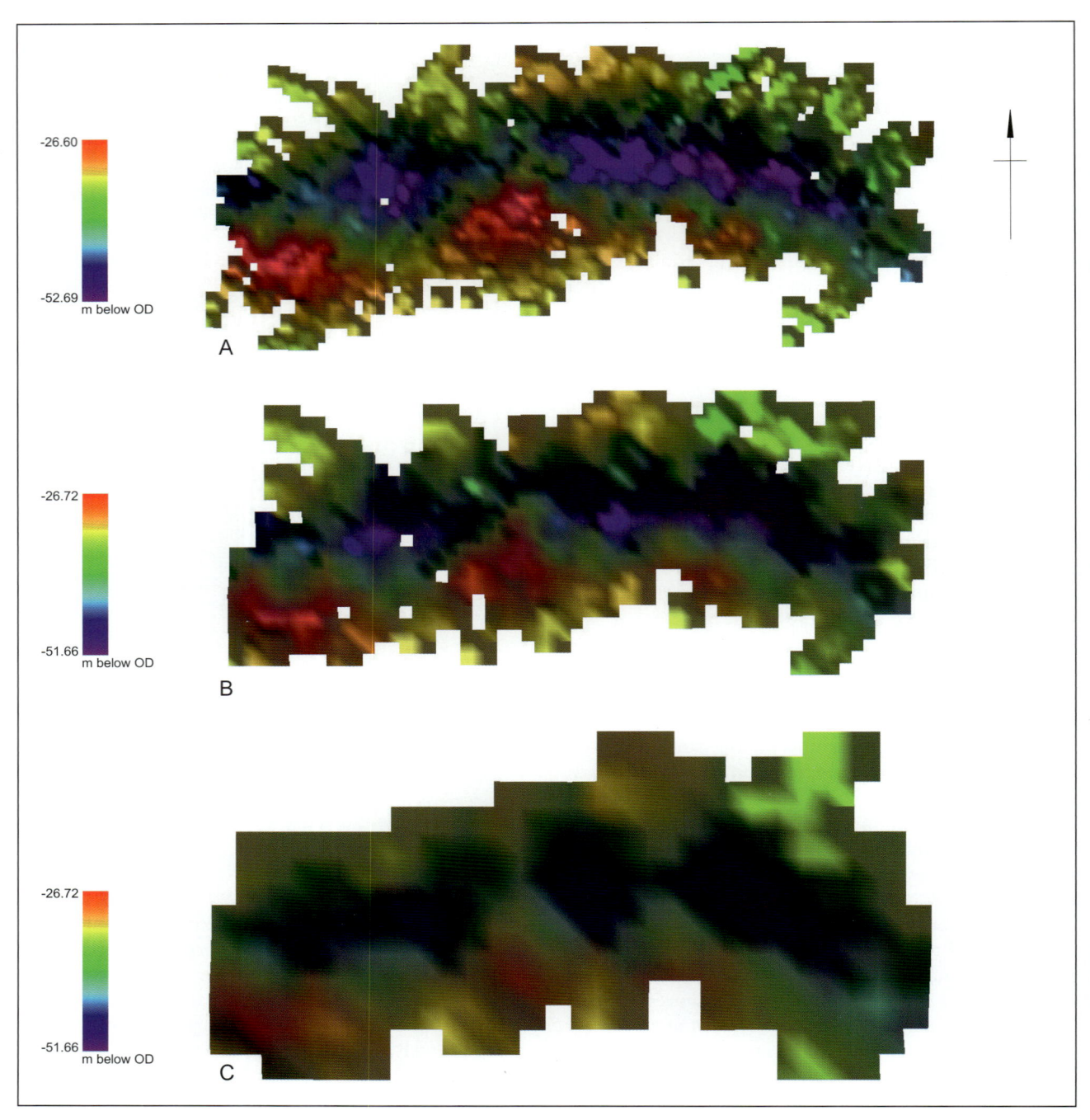

Figure 3.2 Illustration of gridding of data at: A) 50 x 50 m; B) 100 x 100 m; C) 200 x 200 m. The smaller the grid size, the greater the resolution and the greater clarity with smaller features identifiable

scientific beam trawls and high resolution side-scan sonar and further multibeam echosounder data).

A 6 m-long viborcorer was chosen to target specific features identified in the sub-bottom profiles (as being of interest either geologically or archaeologically) and deployed to collect 78 vibrocores from 39 locations across the survey area. At each site, one vibrocore was collected in a clear plastic liner, and a second in a black liner in order to preserve the core for OSL dating. Eight of these vibrocores were situated within the current East Coast dredging block of which VC29, situated 1.4 km west of Area 240, in Area 251, was selected for palaeoenvironmental analysis and dating (Fig. 3.1).

A total of 19 clamshell grab samples were acquired throughout the REC area, one of which CG6, was located in Area 240; this sample targeted the area from which the archaeological material was originally recovered. The clamshell grab successfully recovered a piece of worked flint and this is discussed in more

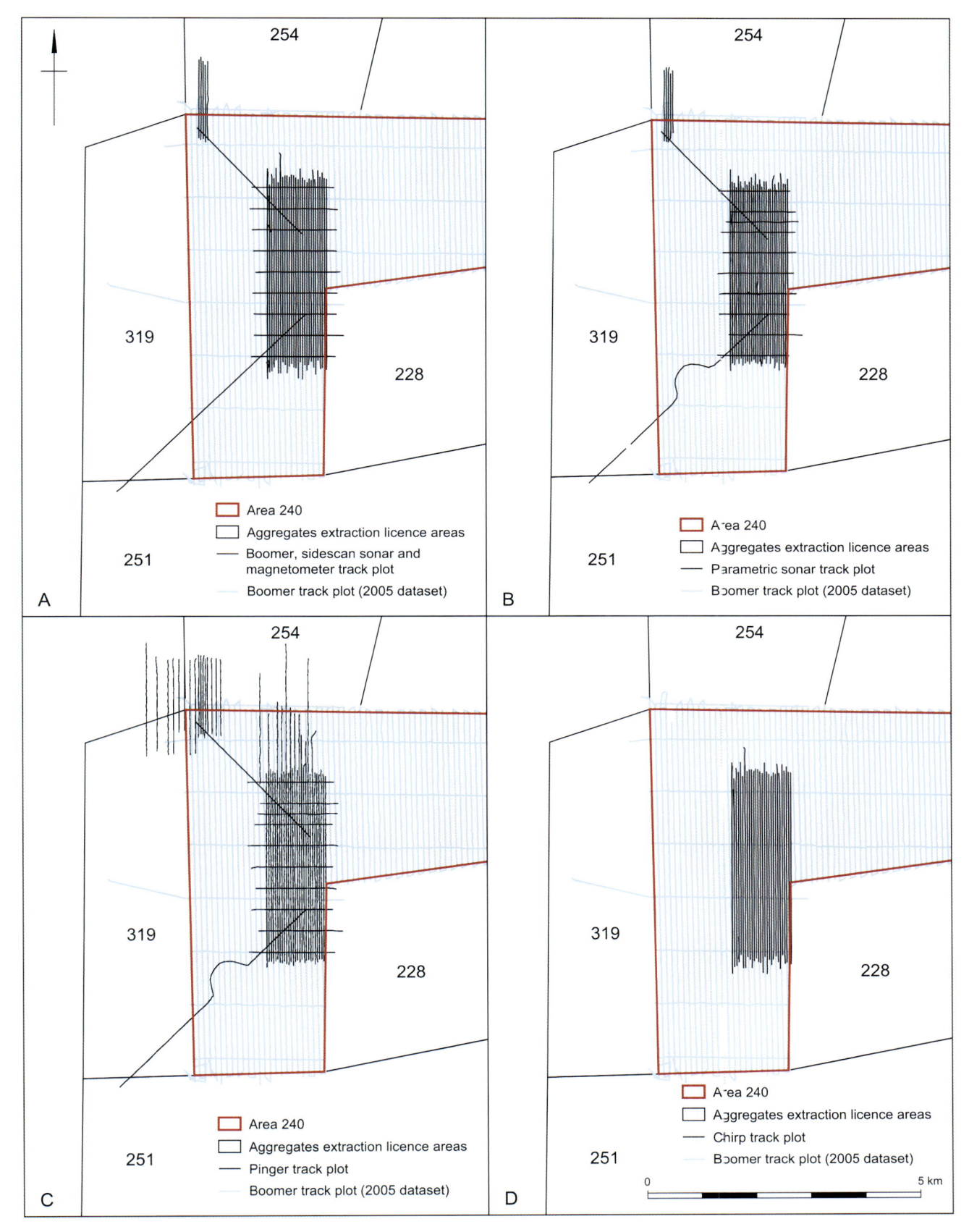

Figure 3.3 Geophysical survey track plots acquired in 2009 for the boomer A) parametric sonar; B); pinger C), and chirp D)

Plate 3.1 Deployment of surface-tow boomer sub-bottom profiler system

detail in Chapter 5. The results of Stage 1 provided a baseline interpretation of the sedimentary units in Area 240 and also, combined with the interpretations from additional surveys, placed the palaeogeography into a more regional context.

Stage 2: Geophysical Data Acquisition

The second stage of the project involved the acquisition and processing of geophysical data within the Archaeological Recovery Zone (ARZ) (Wessex Archaeology 2009b). The aim of the survey was to acquire a more detailed dataset and to assess the different sub-bottom profiler seismic sources for targeting aggregate deposits. The acquisition of the new dataset also established the condition of the area since dredging had ceased (with the implementation of the AEZ) and to ascertain changes in surface and sub-surface sediments compared to the previously acquired 2005 dataset.

The geophysical survey was undertaken by Wessex Archaeology on the MV *Wessex Explorer* between 14 April and 4 May 2009. During the geophysical survey, sidescan sonar, magnetic, single-beam echosounder and four different sub-bottom profiler (boomer, pinger, chirp and parametric sonar) datasets were acquired.

The survey covered approximately 3.5 x 1 km with 28 main lines oriented north–south, (owing to strong, local tidal currents), with a line spacing of 40 m. The ideal spacing for cross-lines would be 80 m, however, due to available survey time and weather conditions on site, up to 10 cross-lines of east–west orientation, and each with a line spacing of either 200 m or 400 m (Fig. 3.3) were acquired.

The main survey area was covered by all sensor types; although no cross lines were acquired with the chirp due to unsuitable weather conditions. There were two line plans for the main survey: one for the sidescan sonar, magnetometer, boomer and chirp, and another survey plan with the survey lines offset by 20 m for the pinger and parametric sonar datasets. Bathymetric data were acquired for all survey lines using the single-beam echosounder and the parametric sonar, respectively. This resulted in a sub-bottom profiler coverage of 20 m line spacing. In accordance with the results of the survey conducted in the Palaeo-Arun, this would result in the identification of small landscape features (Wessex Archaeology 2008a).

Based on the initial interpretation of the 2005 geophysical dataset (Stage 1), additional lines were designed to target specific features: this included two tie-in lines (run to the north-west and south-west of the survey area) and a series of small lines over a specific bank and channel feature (in the north-west corner of Area 240 and Area 254). All datasets, except chirp, were acquired on these additional lines.

Although the survey focused on a smaller total surface area, the 2009 survey was more

Table 3.1 Comparison of 2005 and 2009 sub-bottom profiler and bathymetry datasets

Survey		2005 dataset			2009 dataset	
	Source	Line km	Line spacing	Source	Line km	Line spacing
Sub-bottom profiler	Boomer	360	N–S at 100 m; E–W at 1000 m	Boomer	141	N–S at 40 m; E–W at 200 or 400 m. Plus extra lines
				Pinger	157	N–S at 40 m; E–W at 200 or 400 m. Plus extra lines
				Chirp	108	N–S at 40 m
				Parametric sonar	165	N–S at 40 m; E–W at 200 or 400 m. Plus extra lines
Bathymetry	Multibeam echosounder	360	N–S at 100 m; E–W at 1000 m	Single-beam echosounder	141	N–S at 40 m; E–W at 200 or 400 m. Plus extra lines

comprehensive and the data that was acquired at a much narrower line spacing than the 2005 dataset provided a dataset at a higher lateral resolution (as summarised in Table 3.1).

Positioning

A Trimble GPS in combination with HYPACK survey software was used to provide positioning for the survey. Offsets to towing points and fixed laybacks were used to calculate equipment positioning. A TSS motion reference unit was also used to correct the single-beam echosounder and parametric sonar data for the effects of heave.

Plate 3.2 Deployment of chirp sub-bottom profiler system

Single-beam Echosounder Data

Bathymetry data were collected to provide information on water depth and seabed morphology, and to generate a vertical reference datum for the identified horizons from the sub-bottom profiler data. Data were also used to compare water depths to the 2005 dataset in order to establish any bedform movement and assess the effects of dredging since 2005. The single-beam echosounder data were acquired using a Knudsen 320M dual frequency echosounder working at 3.5 and 250 kHz simultaneously. A DMS3-05 motion reference unit was used to correct the data for heave and tidal variation, which was applied during post-processing.

Plate 3.3 On board recording of parametric sonar data

Sub-bottom Profiler Data

Four sub-bottom profilers were trialled: boomer (Pl. 3.1), pinger, chirp (Pl. 3.2) and parametric sonar (Pl. 3.3, Table 3.2). The acquisition of the four datasets allowed an assessment to confirm which of the sub-bottom profilers achieved best results regarding data quality, resolution, penetration and applicability for the sediment type (sands and gravels) found at the site and which could be further recommended, specifically for archaeological applications, in similar environments.

Table 3.2 Details of sub-bottom profiler systems used in the Stage 2 geophysical survey

Sub-bottom profiler	Model	Dominant frequency	Maximum vertical resolution	Ping rate in Hz	Beam width	Radius in m (in x m of water)	Sub-seabed penetration (m)	Towing depth
Boomer	Applied Acoustics Model AA200	1 kHz	0.4 m	3	30–40°	17.5 m (in 25m; for 35°)	30	sea surface
Pinger	GeoAcoustics 136A sub-tow pinger; hull mounted pinger	3.5 kHz	0.1 m	5	55°	35.7 m (in 25 m)	6	sea surface
Chirp	Edgetech SB216S chirp system	2–15 kHz	0.06 m	6	17°	5.2 m (in 17 m)	2	8 m below sea surface
Parametric sonar	SES2000 compact	8 kHz	0.05 m	~15	1.8°	0.8 m (in 24 m)	6	1 m below sea surface; on pole

> **Box 3.3: Sidescan sonar**
>
> The sidescan sonar can be used to assess features, both natural and anthropogenic, on the seabed. The system is generally towed on a cable behind the survey vessel and consists of transducers on either side of a towfish which emit pulses of acoustic energy in the direction perpendicular to travel. The acoustic energy is reflected from the seafloor to the transducers and the strength of the pulses recorded via a workstation on board the vessel.
>
> The strength of the reflections is dependent on the properties of the seafloor material; different sediment types produce different strength signals. This results in an acoustic image of the seabed relief and indicates the character of the seabed sediments. Sidescan sonar systems are commonly used to assess upstanding man-made materials such as wrecks (ship and aviation), debris, etc.

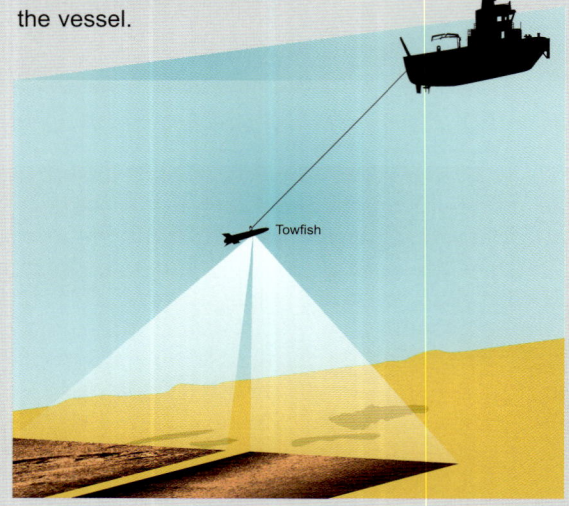

The chirp seismic source generates a sweep over a user-defined frequency range and the reflected signal is passed through a pulse compression filter. It can achieve high resolution, but comparatively low penetration (eg, the typical penetration in calcareous sand is 6 m). The chirp dataset was generally of moderate to low quality; penetration was approximately 2 m to 3 m and vertical resolution was much lower than expected. Because the chirp was towed at approximately 8 m below the sea surface, the data was less affected by weather and the radius of the footprint is relatively small at 5.2 m. The comparatively low data quality is thought to be mainly due to the sediment types present at site. Processing of these datasets followed the same protocols as the Stage 1 interpretation, as described above.

The parametric sonar seismic source functions on principles different than other, more commonly used sub-bottom profilers (as described in Box 3.1). Parametric arrays generate their acoustic signal using non-linear principles: if two close frequencies are transmitted at very high levels, non-linear propagation of the sound signal induces a secondary wave with a frequency equal to the difference between the two primary frequencies (Lurton 2002). By using the high frequency to generate the low frequency, very small beam widths are achieved; also, the ping rate is considerably higher. Furthermore, the parametric sonar is pole-mounted and thus less susceptible to adverse weather conditions than surface-towed systems. Due to relatively high secondary frequencies, high resolution is achieved; however, penetration is comparatively low in sediments such as sand and gravel.

The parametric sonar dataset was acquired using a SES2000 compact system mounted on a pole over the starboard side at 1 m below the sea surface. The parametric sonar was deployed used a primary frequency of 100 kHz with an adjustable secondary frequency (between 5 to 15 kHz); in this way, the instrument acts as a single-beam echosounder as well as a sub-bottom profiler. Penetration of up to 6 m was achieved and resolution was generally very good (up to approximately 0.1 m resolution). The parametric sonar dataset allowed for accurate mapping of the thickness of the Holocene sediments as it was of the highest vertical resolution and was not affected by any artificial effects, such as ringing. Horizontal resolution was excellent due to high ping rates of approximately 15 Hz which equates to a reading at 0.13 m at a typical survey speed of four knots. Beam width was extremely low; for the specified beam angle of 1.8°, the radius of the footprint is only 0.8 m. This means that lateral resolution is extremely good, but also that only reflectors directly below the instrument are shown – reflectors just outside the footprint are not imaged.

The boomer seismic source operates at high power and comparatively low frequencies of approximately 1 kHz. Thus, it achieves greater penetration, but relatively low resolution. Boomers are very well suited to sand and gravel sediments (Parkinson 2001) and are routinely used for site investigation work by marine aggregate companies in UK waters. Throughout the survey the boomer data were of generally good quality with sub-seabed penetration in excess of 30 m.

The pinger seismic source operates at higher frequency, typically 3.5 kHz, and with lower power than the boomer. The system achieves higher resolution, but much lower penetration than the boomer seismic source. Generally, the quality of pinger data is adversely affected by certain sediment types, especially gravels and gravelly sands. In this survey sub-seabed penetration was limited to 6 m.

Sidescan Sonar Data

Sidescan sonar data were acquired to provide information on the seafloor sediments and bedforms, recent dredging and trawling activity in the area, as well as anthropogenic features such as wrecks and debris, etc. The data were acquired using a Klein System 3000 dual frequency sidescan sonar towfish operating at 500 and 100 kHz and a range of 50 m for the main lines and 50/150 m for the cross lines. The data were recorded in SonarPro version 11.3 as xtf files and were processed using Coda Geosurvey software.

Magnetometer Data

The magnetometer was used to identify magnetised sediments within the area and to identify magnetic signal caused by anthropogenic features on, or buried beneath, the seabed such as wrecks, debris, etc. The magnetometer data were acquired using a Marine Magnetics Explorer magnetometer which was cycled at 4 Hz; accuracy is 0.2 nT and was processed using MagPick software.

Box 3.4: Magnetometer system

Marine magnetometers can be used to detect ferrous material lying on or buried below the seabed. They are also used to assess variations is magnetised sediments on the seabed through detecting alterations in the strength of the earth's magnetic field. The magnetometer towfish is towed behind the survey vessel individually or piggy-backed off the sidescan sonar towfish at a sufficient distance not to detect the survey vessel.

Methodology Discussion

The interpretation focused on the sub-bottom profiler data, supplemented by the other datasets. The geology of the area had the most significant impact on the quality and usefulness of the data in terms of interpreting sediment units and features sub-seabed.

The boomer data proved to be the most useful of the datasets for interpretation at this site. Both penetration (between 16 and 35 m sub-seabed) and resolution were sufficiently high to identify geological units, such as tops and bases of formations, as well as small-scale features such as bases of sandwaves, shallow cuts and fills and erosional features. It was also possible to determine which geological units are present at the surface. The parametric sonar had limited penetration (up to a maximum of 6 m) but had a higher vertical resolution than the boomer system. The parametric sonar resolved the thickness of the uppermost sediments, the bases and internal layering of sand waves as well as resolving shallow cuts, fills and erosional features where penetration permitted.

Both the pinger and chirp systems were strongly affected by the nature of the sediments (and bedforms) and were found to yield the least useful datasets for interpretation in this area. Unlike the data acquired in Area 254 where pinger data was successfully used to delineate fine-grained sediment units, the coarser-grained sediments encountered in Area 240 prevented penetration further than a maximum of 6 m and caused pronounced ringing in the data obscuring any small-scale features such as cuts, fills and erosional surfaces. Interpretation was further complicated by the diffraction of the seismic energy by small sand ripples which are present over most of the area.

Also, the chirp system, which proved successful in the *Seabed Prehistory: Happisburgh and Pakefield* project targeting similar sediment types (Wessex Archaeology 2008f), failed to reproduce its effectiveness during this survey due to the nature of the surface sediments. Penetration was limited to the upper 3 m sub-seabed and resolution was much lower than expected. The parametric sonar, which had not previously been used for this type of archaeological survey in the UK proved to be the most useful system for providing additional information for the interpretation of the boomer data (eg, by resolving shallow features and the thickness of Holocene sediments). Therefore, as with previous survey in the area, this project proved the effectiveness of the application of multiple types of sub-bottom profiler at a single location.

Comparison and Integration of 2005 and 2009 Datasets

The sediment units identified during the 2005 and 2009 datasets were comparable, but differences in interpretation were apparent. Although the data quality of the 2005 dataset was better due to more favourable weather conditions during acquisition, the denser line spacing and use of multiple sources resulted in a higher resolution interpretation from the 2009 datasets. Additionally, the parametric sonar provided a more detailed view of the upper few metres than the 2005 boomer dataset, allowing a more detailed interpretation of the structure within the uppermost sediment unit of the ARZ.

The complexity of sediment units within the area, especially with regards to mapping faint and shallow reflectors, calls for data of the highest possible quality and resolution. Comparison with the 2005 dataset showed that weather conditions during data acquisition in 2009 play a crucial role. Although allowance was made for weather downtime in the survey design of the 2009 survey, it was not practical to wait for perfectly calm seas; therefore most datasets were subject to some degradation due to the marginal weather conditions, and the high-frequency sub-bottom profiler data were particularly affected.

An initial comparison of the four sub-bottom datasets acquired in 2009 resulted in the use of the parametric sonar and boomer data as these were the most useful for archaeological interpretation. This was supplemented where appropriate with the pinger and chirp data. The interpretations of the Stage 1 and 2 assessments were then combined, which resulted in a higher resolution interpretation in the ARZ. Although the datasets were acquired four years apart and there were obvious differences in both the nature of the seabed and the vertical extents of the units (due to dredging activities in the intervening years), the lateral extents of the units were comparable and an integrated interpretation was produced.

The processed and modelled bathymetry data from the 2009 survey was compared to the 2005 dataset in order to assess the changes in the seabed morphology. Although subtle differences in the morphology were observed, there were difficulties in assessing whether the cause was due to natural hydrodynamic influences or due to the effects of dredging. However, spot checks in areas without sandwaves have shown on average 1.0 to 1.5 m difference in depths between 2005 and 2009; this has been interpreted to indicate removal of sediment in parts of the site within that time period.

Stage 3: Seabed Sampling

Stage 3 involved sampling the seabed for artefacts. This was the focus of the second strand of research, namely to establish the presence of any remaining archaeological material since the original discovery. Methods involved the trialling of remote sampling techniques including stills photography, scientific trawling and clamshell grab sampling along specified transects targeting specific sediment unit types (detailed methodology and results are discussed in Chapter 5). Although it was not the aim of the seabed-sampling survey to inform the palaeogeographic reconstruction of Area 240, details on the seabed sediments recovered during sampling furthered the understanding of the uppermost sediment unit.

Stage 4: Palaeoenvironmental Sampling

Stage 4 involved the acquisition of vibrocores at 10 locations within the Area 240 (Fig. 3.4). The vibrocores targeted sediment units that would provide palaeoenvironmental material and allow for the assessment, dating and reconstruction of the former land surfaces within Area 240. Assessments were based on the on-going interpretation of geophysical and geotechnical data (Wessex Archaeology 2010a).

The vibrocore survey took place between the 10 and 13 May 2010 on board the *VOS Baltic*, a 61 m long offshore support vessel (Pl. 3.4). The survey was sub-contracted to Fugro Alluvial Offshore Limited as part of a joint coring program between Wessex

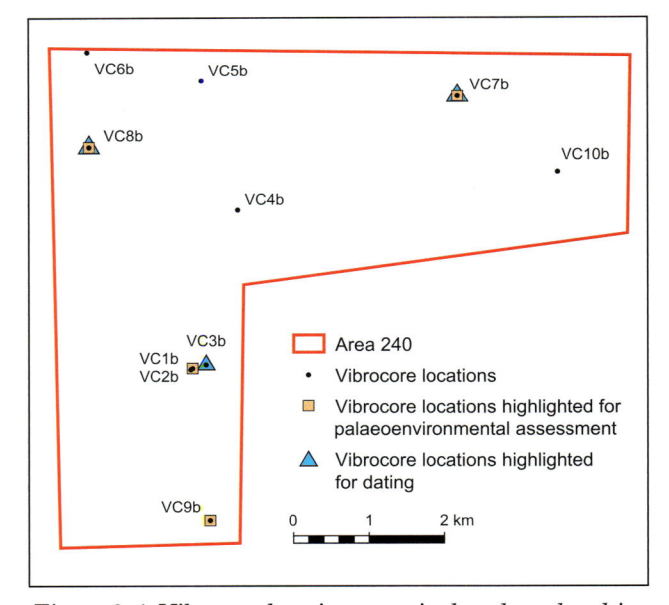

Figure 3.4 Vibrocore locations acquired and analysed in 2009. Four vibrocores underwent palaeoenvironmental assessment for macro-botanical (plants and charcoal), macro-fauna (molluscs and insects), micro-botanical (pollen, diatoms and microcharcoal) and micro-fauna (ostracods and foraminifera). A series of samples were dated using radiocarbon and OSL dating techniques

Archaeology and the National Oceanographic Centre. The acquisition of the vibrocores was conducted by Fugro Alluvial Offshore Limited; the positioning and the recording of water depths at each location were conducted by Fugro Survey Limited.

Two vibrocores were acquired at each location. The first was acquired in a clear liner to be used for geoarchaeological logging (with suffix c, eg, VC2c) and a second in an opaque liner to be used for OSL dating (suffix b, eg, VC2b). All vibrocores were acquired with the exception of the OSL core at location VC8 (due to strong tidal currents). The positions of the vibrocores were recorded using DGPS with typical positioning accuracy of 0.5 m. An echosounder was provided to ensure accurate depth measurements at each vibrocore location. Water depths were corrected for vessel draft and reduced relative to Chart Datum (CD) using predicted tides (extrapolated to Area 240 from the nearest Standard Port of Lowestoft). These were then subsequently converted to Ordnance Datum (OD).

A High Performance Corer was used to acquire the vibrocores. The design utilises innovative electric motor and sample barrel design allowing successful acquisition of a range of sediments, including gravels and sands. The samples were acquired with a 6 m barrel and recovery ranged between 2.81 m and 6.00 m. The diameter of each core was 100 mm. From the 19 vibrocores in excess of 97 m of sediment were recovered.

Onshore, the vibrocores acquired in clear liners, were split longitudinally using an angle grinder and carefully prised open with a knife in order to preserve sedimentary structure. One half of each core was cleaned where necessary, and photographed. Each vibrocore was then geoarchaeologically recorded providing details of the depth to each sediment horizon and the character of the sediment (Pl. 3.5). Sedimentary characteristics were recorded using standard practice that includes documenting texture, colour, stoniness and depositional structure (*cf.* Hodgson 1976).

Details from the vibrocore logs were compared to the geophysical data (2005 and 2009) in order to characterise the geophysical units identified at the core locations and to extrapolate the interpretation for Area 240. Although the geophysical data and sediment descriptions can be compared, which is useful for broad-scale ground-truthing of the geophysical data, there may be differences in depths of horizons from each data type and subtle changes in density may be obvious in the geophysical data but not the sediment data. Equally, subtle changes in the sediment type do not necessarily result in a density change significant enough to be recorded in the sub-bottom profiler data. Also, the resolution of the sub-bottom data can mean that thinner sediment units are not resolved. The water depths at the locations of the vibrocores also provided spot depths which were useful for comparing variations in recorded depth between 2005, 2009 and 2010. Based on the integrated interpretation, a series of vibrocores were selected for sub-sampling for the purpose of palaeoenvironmental assessment, analysis and dating.

Stages 5 and 6: Palaeoenvironmental Assessment, Analysis and Dating

Samples from four vibrocores (VC2c, VC7c, VC8c1 and VC9c, Fig. 3.5) were assessed for macro-botanical remains (plants and charcoal), macro-fauna (molluscs and insects), micro-botanical remains (pollen, diatoms and micro-charcoal) and micro-fauna (ostracods and foraminifera). Additionally, a series of samples were dated using radiocarbon

Plate 3.4 Deployment of the High Performance Vibrocorer from the vessel Vos Baltic. *A number of 6 m long cores were acquired (© Wessex Archaeology and NOC)*

Plate 3.5 Geoarchaeological recording of a vibrocore

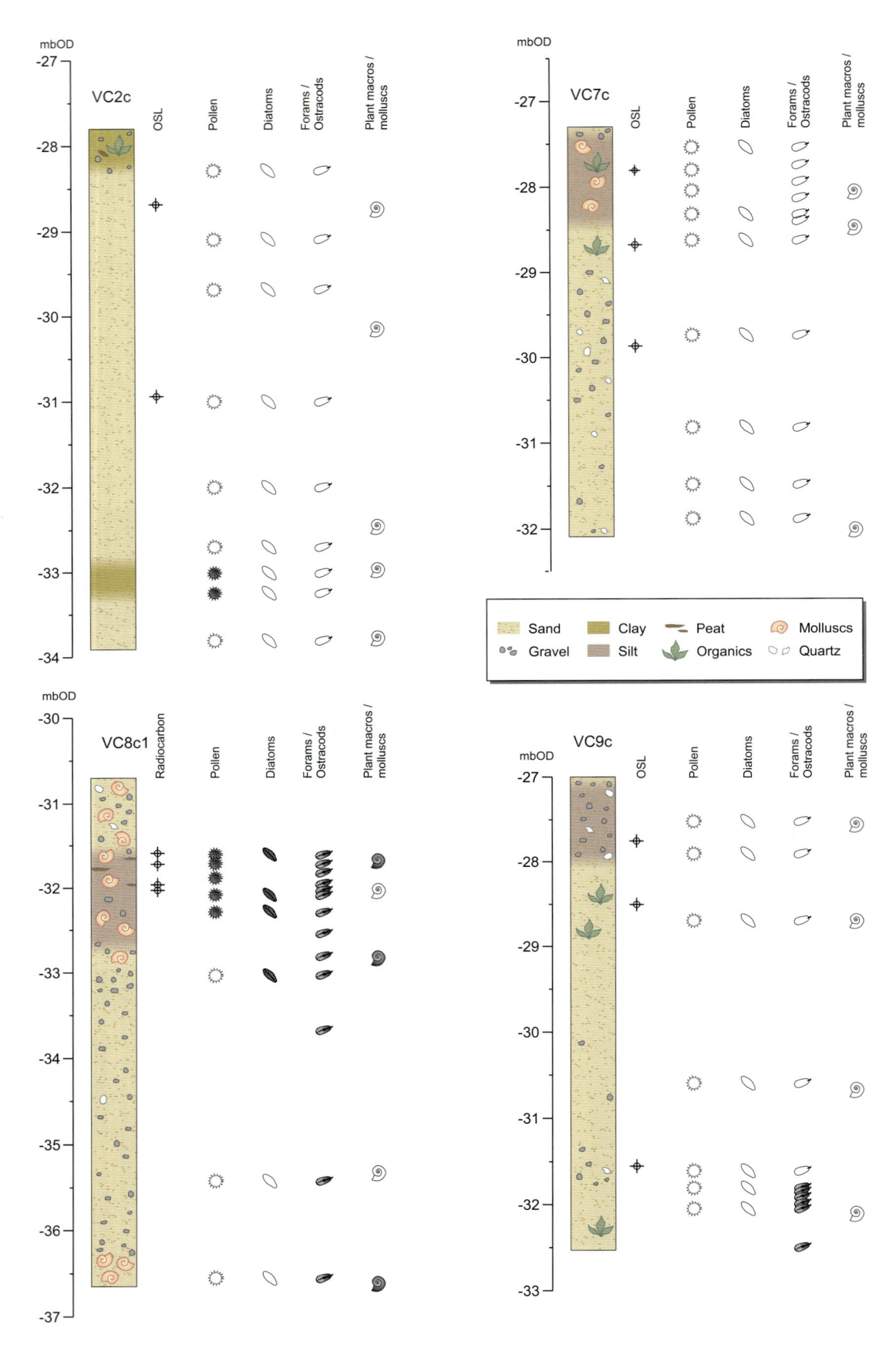

Figure 3.5 Palaeoenvironmental sampling of vibrocores VC2c, VC7c, VC8c1 and VC9c. Infilled symbols indicate that the sample was productive; clear symbols indicate barren samples

(VC8c1) and OSL dating (VC3b, 7b and 9b) techniques.

Mollusca, waterlogged plant, charcoal and insect assessments were carried out on 18 samples (Pl. 3.6). Samples of approximately 100 to 250 ml were processed by wet-sieving using a 0.25 mm mesh size. The samples were visually inspected under a x10 to x40 stereo-binocular microscope to determine if waterlogged plant remains were preserved. Where molluscs were present, preliminary identifications and quantifications of dominant taxa were conducted. Habitat preferences follow those described by Barrett and Yonge (1958) and Kerney (1999).

Specialist palynological assessment was used to provide information on past flora and environments. When a stratified sequence of sediment is investigated, pollen analysis can show how the pollen arriving at the site of deposition has varied over a given period of time, and therefore allow interpretations relating to climate change, vegetation history, human activity and the modification of the local environment. Pollen can be preserved in a range of environments, but preservation is principally determined by whether they are anoxic, such as sediments deposited in lakes, fens, mires and buried soils. This project employed standard preparation procedures (Moore *et al.* 1991): 4 cm^3 of sediment was sampled, with a *Lycopodium* spike added to allow the calculation of pollen concentrations. All samples received the following treatment: 20 ml of 10% KOH (80°C for 30 minutes); 20 ml of 60% HF (80°C for 120 minutes); 15 ml of acetolysis mix (80°C for 3 minutes); stained in 0.2% aqueous solution of safranin and mounted in silicone oil following dehydration with tert-butyl alcohol. Pollen counting was conducted at a magnification of x400 using a Nikon Eclipse E400/Nikon SE transmitted light microscope. Determinable pollen and spore types were identified to the lowest possible taxonomic level. The pollen and spore types used are those defined by Bennett (1994; Bennett *et al.* 1994), with plant nomenclature ordered according to Stace (1997).

Microscopic charcoal analysis was undertaken using the original pollen preparation residues and the method used was adapted from Clark (1982). The amount of charcoal present was quantified by counting a minimum of 200 random fields of view, applied to each slide, and the number of charcoal particles recorded. A minimum of 50 *Lycopodium* spores, added as part of the original preparation procedure, were also counted, thus enabling the calculation of charcoal concentrations.

Diatom assessment was carried out by Dr Nigel Cameron at the Environmental Change Research Centre, University College London. The diatom assessment of each sample takes into account the numbers of diatoms, the state of preservation of the

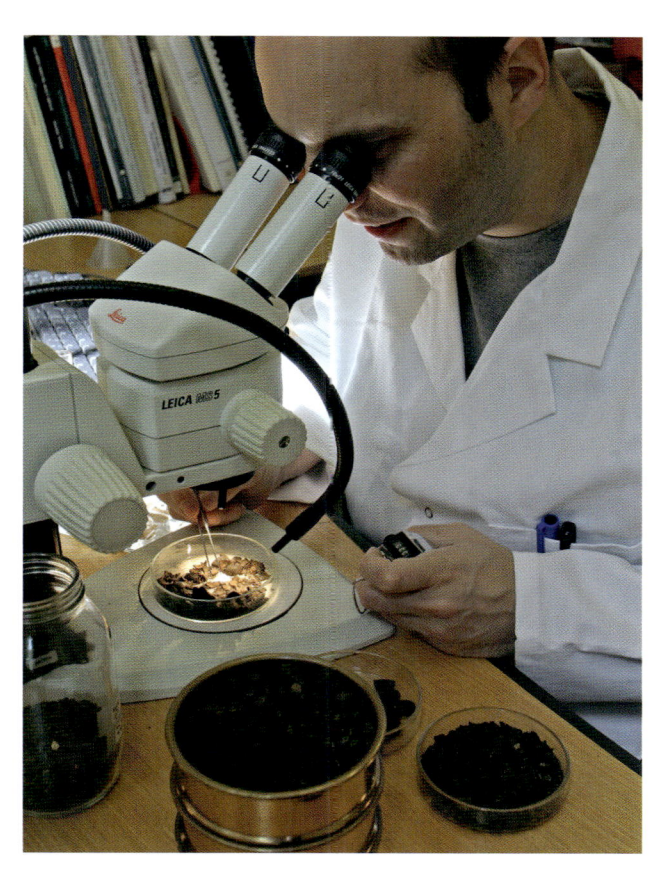

Plate 3.6 Assessing and counting plant remains

diatom assemblages, species diversity and diatom species environmental preferences. Diatom preparation followed standard techniques (Battarbee 1986; Battarbee *et al.* 2001). Diatom floras and taxonomic publications were consulted to assist with diatom identification; these include Hendley (1964), van der Werff and Huls (1957–1974); Hartley *et al.* (1996) and Krammer and Lange-Bertalot (1986–1991). Diatom species' salinity preferences are discussed in part using the classification data in Denys (1992), Vos and de Wolf (1988; 1993) and the halobian groups of Hustedt (1953; 1957).

For the ostracod and foraminifera sediment samples of approximately 25 g were disaggregated in a weak solution of hydrogen peroxide and water, then wet sieved through a 63 µm sieve. The sediment was dried and sieved through 500 µm, 250 µm, 125 µm sieves. Microfossils were picked out under 10–60 x magnification and transmitted and incident light using a Vickers binocular microscope. Where possible a minimum of 100 specimens per sample were picked out and kept in card slides. Identification and environmental interpretation of ostracods follows that of Athersuch *et al.* (1989) and Meisch (2000) and foraminifera (Murray 1979; 1991).

Four radiocarbon samples from a single sediment unit in VC8c1 were submitted for dating. The samples were submitted by English Heritage for Accelerator

Mass Spectrometry (AMS) radiocarbon dating to the Oxford Radiocarbon Accelerator Unit at Oxford University and the Scottish Universities Environmental Research Centre (SUERC) in East Kilbride. Unfortunately both of the samples sent to Oxford failed due to low yields (Brock et al. 2010 describe relevant pre-treatments). The samples sent to SUERC were successfully dated, following technical procedures described by Vandenputte et al. (1996); Slota et al. (1987); and Xu et al. (2004). Internal quality assurance procedures and international inter-comparisons (Scott 2003) indicate no laboratory offsets, and validate the measurement precision given. The resulting dates yielded were reported as conventional radiocarbon years before present, or BP (Stuiver and Polach 1977), and quoted according to the format known as the Trondheim convention (Stuiver and Kra 1986). All dates were then calibrated against the IntCal09 Northern Hemisphere radiocarbon curve (Reimer et al. 2009) using the program OxCal 4.1 (Bronk Ramsey 1995; 2001). Calibrated dates are quoted as calendar years BC, with date ranges quoted using the 2σ calibrated range (95.4%) and end point rounded outwards to 10 years in the form recommended by Mook (1986). Dates older than 15 ka ^{14}C BP are rounded to the nearest 50 years following the data spacing of the IntCal09 dataset (Reimer et al. 2009).

Nine samples were submitted for OSL analysis. Dating was carried out by Dr Phil Toms at the Geochronology Laboratories, University of Gloucestershire. The time-dependent optically stimulated luminescence signal was calibrated from multi-grain, fine sand aliquots using a single-aliquot, regenerative-dose protocol to provide a measure of natural dose absorption during the burial period. This dosimetry was converted into chronometry by assessing the rate of dose absorption, accounting for litho-cosmogenic emissions along with moisture absorption and grain size attenuation effects (Toms 2011).

In light of the sedimentary and palaeo-environmental analysis the geophysical data interpretation was re-addressed and an integrated interpretation produced.

Subsequent Work

The Area 240 project was complete by March 2011 and the results of the project synthesised into a final project report (Wessex Archaeology 2011). However, work on the palaeogeography of the region has since continued. In 2012 Wessex Archaeology was commissioned by the British Marine Aggregate Producers Association on behalf of The Crown Estate, CEMEX UK Marine Limited, Hanson Aggregates Marine Limited, Lafarge Tarmac Limited and Volker Dredging Limited (the East Coast aggregate block licensees), to conduct an assessment of the Palaeo-Yare catchment area, East Anglia. The aim of the project was to delineate at a catchment level, where possible, the extents and survival of the specific sediment units from which a large number of flint artefacts and faunal remains were recovered in Area 240 (Wessex Archaeology 2012).

The importance of the recovery of artefacts from Area 240 was acknowledged by the aggregate industry and that the archaeological importance of the region needed to be assessed in order to understand the needs of mitigation and monitoring for future licence applications and operations. The project was conceived in order to allow the development of a regional framework which would result in a better understanding of the prehistoric archaeological resource in the region in terms of its distribution, significance and the mitigation effects from dredging.

Approximately 2500 line kilometres of sub-bottom profiler data (from 22 surveys) and 1171 vibrocore logs (from 43 separate surveys acquired between 1988 and 2011), were reviewed. The majority of these logs were from sampling originally undertaken by the marine aggregate industry (Fig. 3.6). Additionally, approximately 400 onshore borehole logs (supplied by BGS) were reviewed, providing further context.

Although this volume focuses on the work carried out in Area 240 that predates the catchment-level work, the overarching results of the Palaeo-Yare catchment assessment project are incorporated into the discussions on the development of the regional palaeogeography, and in particular, the development of the Palaeo-Yare Valley system.

Summary

As discussed in the introduction to this chapter, scale is an important factor when reconstructing landscapes. The work carried out in Area 240 has provided an in-depth study of the palaeogeography in what is a relatively small area of seabed. The geophysical data (Stage 1 and 2) allowed for an interpretation of geological structures in the area; these have since been ground-truthed through the geotechnical integration (Stages 1 and 4). The palaeoenvironmental analysis (Stage 5 and 6) provided added detail on the likely environments at different times throughout the development of Area 240 and the dating provided chronological context. Of course, more palaeoenvironmental analysis and more dating may further support the interpretation, however based on the all the available data analysed, the resulting reconstruction of Area 240 has allowed

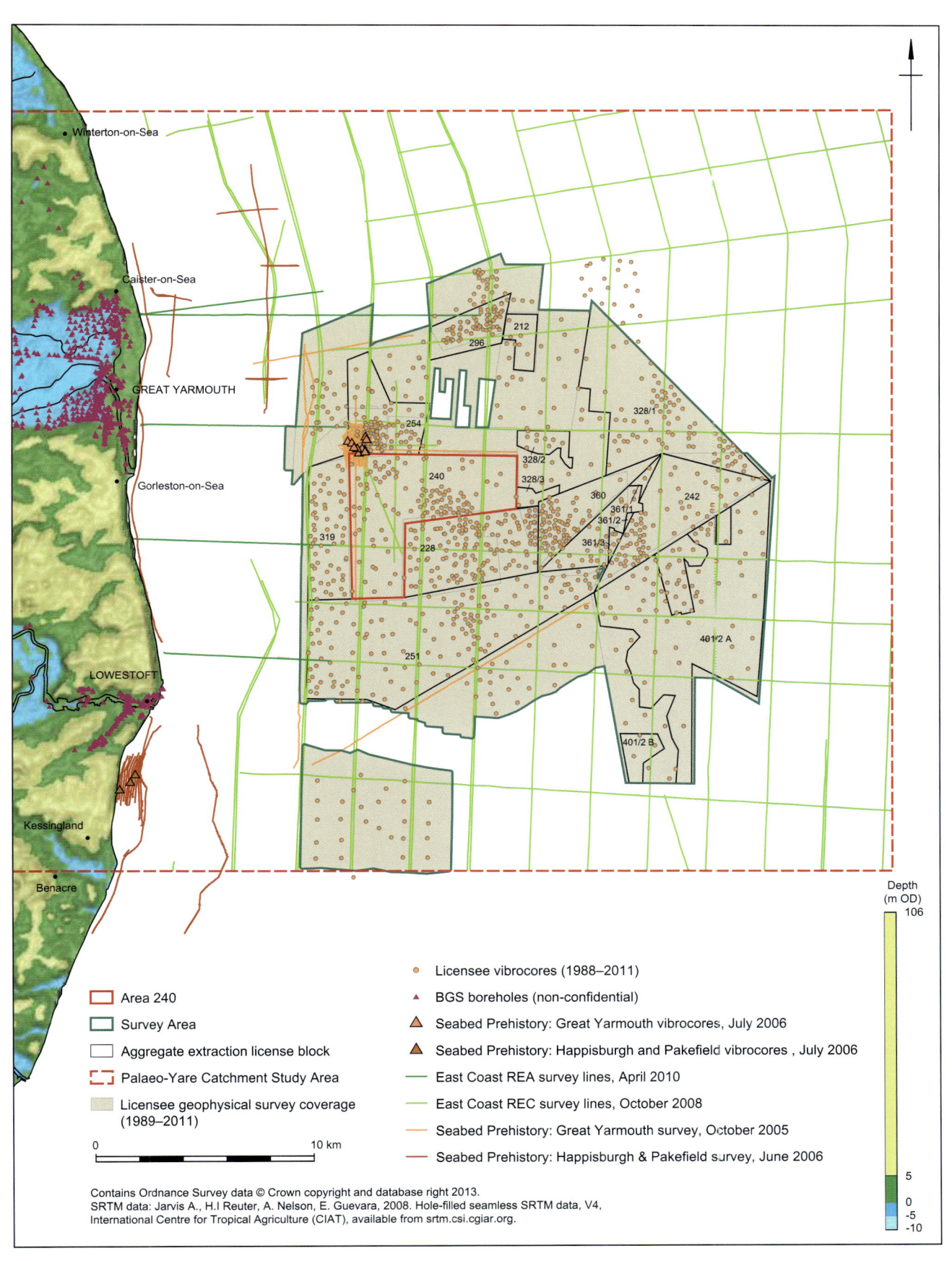

Figure 3.6 Overview of marine geophysical and marine and terrestrial geotechnical datasets acquired for assessment of the Palaeo-Yare Valley

for hypotheses to be made regarding the context and source of the archaeological material. This reconstruction alone, however, could not provide a broader picture of how Area 240 related to the development of the region.

Although Area 240 is situated in 20–35 m water depth, the site is interpreted to exist in what is a now-submerged section of the Palaeo-Yare river system. As such, it cannot be evaluated in geographic isolation. Although relatively little work has been done on the onshore Yare Valley, the river infill deposits and river terraces, where known, have been documented (Arthurton *et al.* 1994; Moorlock *et al.* 2000). The recent work carried out was designed to review the Palaeo-Yare Valley at catchment level, linking the present-day onshore with the offshore environment. The assessment of geotechnical and geophysical data in the aggregate block has allowed the Area 240 interpretation to be expanded to the immediate region and increased the confidence in the initial interpretation.

The Palaeo-Yare catchment interpretation is, in turn, placed in a regional context by the results of the East Coast REC study. These increasing spatial scales have allowed the palaeogeography (and therefore the recovered archaeological material) of Area 240 to be situated within a broader context, the results of which are presented in the following chapter.

Chapter 4
Prehistoric Characterisation of Area 240 and the Southern North Sea Region

Introduction

Over recent decades there has been a tremendous amount of research undertaken in East Anglia with the main focus on the pre-Anglian river systems of the Ancaster, Bytham (previously known as the Ingham) and Thames (eg, Rose *et al.* 2001; 2002; Rose 2009; Westaway 2009; and for discussion of the geological evidence see Gibbard *et al.* 2012) and associated occupation sites at Pakefield, Suffolk and Happisburgh, Norfolk relating to the earliest occupation of Britain (Parfitt *et al.* 2005; 2010; Ashton *et al.* 2014). This area is also a type site for the Anglian deposits (tills) and has been extensively studied (eg, Read *et al.* 2007; Pawley *et al.* 2008; Lee 2009). However, there has been relatively little study on the post-Anglian drainage systems developed at the end of the Anglian cutting into the underlying till and crag deposits. Work carried out has either been regional geological studies (Arthurton *et al.* 1994; Moorlock *et al.* 2000), environmental studies (Coxon 1979) with very few localised studies (eg, as detailed in Rose 2009). Prior to this research, offshore work on the Palaeo-Yare had been confined to Area 254 (Bellamy 1998; Wessex Archaeology 2008c).

There is a rich archaeological record particularly for Lower Palaeolithic sites and artefacts in the region (Wymer 1999; Pettit and White 2012). Key sites such as Happisburgh (Parfitt *et al.* 2010) and Pakefield (Parfitt *et al.* 2005) in coastal positions have provided important artefactual and palaeoenvironmental records. Investigations at Happisburgh have also revealed the oldest known hominin footprint surface outside Africa at between approximately 1 million and 0.78 million years ago (Ashton *et al.* 2014). Similarly inland sites such as Hoxne, Suffolk (Singer *et al.* 1993; Ashton *et al.* 2008) and Norton Subcourse, Norfolk (Schreve 2004) preserve important artefact assemblages and palaeolandscape archives, respectively. These archaeological sites are closely related to their contemporary drainage configuration and glacial history of the region requiring any analysis of the earliest archaeological record to be contextualised by the changing pattern of pre- and post-Anglian river systems and topography.

There exists a strong relationship between landscape (fluvial development and glacial history) and archaeology within the Palaeo-Yare Valley (Wessex Archaeology 2013b). Large artefact assemblages of post-Anglian Lower Palaeolithic tools have been recovered, particularly in the vicinity of Norwich such as Keswick Mill Pit (Sainty 1933) and Whitlingham (Sainty 1927), which are associated with the preserved river terraces. There are also a significant number of reported Palaeolithic artefacts archived within the regional Historic Environment Records (HER); though the majority of these lack primary context (Wymer 1999; Wessex Archaeology 2013b). Wymer (1999) has noted that the gravels in the Rivers Yare and Wensum were deposited under high energy glacial outwash conditions at the end of MIS 8 (*c.* 250 ka); perhaps contributing to a paucity of *in situ* evidence. The Lower Palaeolithic sites of the Palaeo-Yare are reported as being particularly rich in implements, with excavations at Whitlingham producing several hundred lithic artefacts (Sainty 1927; 1933; Wymer 1999). Dating is poorly constrained for these sites, based on artefact typology and river terrace stratigraphy where they are associated. The Lower Palaeolithic hand axe assemblage from Whitlingham is tentatively attributed to around MIS 11–9, 424–300 ka, (Pettit and White 2012). The assemblages from Keswick Mill Pit and Carrow Road, Norwich may be younger, reported to incorporate elements of Middle Palaeolithic Levallois technology (Wymer 1999). There are very low numbers of reported Levallois artefacts within the Palaeo-Yare catchment suggesting some MIS 9–8 hominin activity. Flakes and a core from Keswick Mill Pit (Wymer 1985; 1999), a flake from Carrow Road, Norwich (Sainty 1933) and three reported Levallois flakes from Lenwade Pits at Great Witchingham on the River Wensum (Wymer 1985; 1999) are reportedly associated with hand axe assemblages. Carrow Road being speculatively attributed to MIS 8 (300–243 ka) (Wymer 1999) and Lenwade Pits being attributed to MIS 7 or younger (<191 ka). These assemblages suggest some evidence for <MIS 9 (possibly MIS 8) activity in the upper Palaeo-Yare catchment but the artefactual evidence and absolute dating is currently quite sparse.

Regionally, Lynford Quarry MIS 4–3 (65–57 ka) assemblage represents the first Late Middle Palaeolithic site following the hiatus of human activity between MIS 6 and 3 (57–19 ka) (Boismier *et al.* 2003, Boismier *et al.* 2012). Later Upper Palaeolithic and Mesolithic artefacts are relatively common within the Palaeo-Yare catchment; lithic working sites are

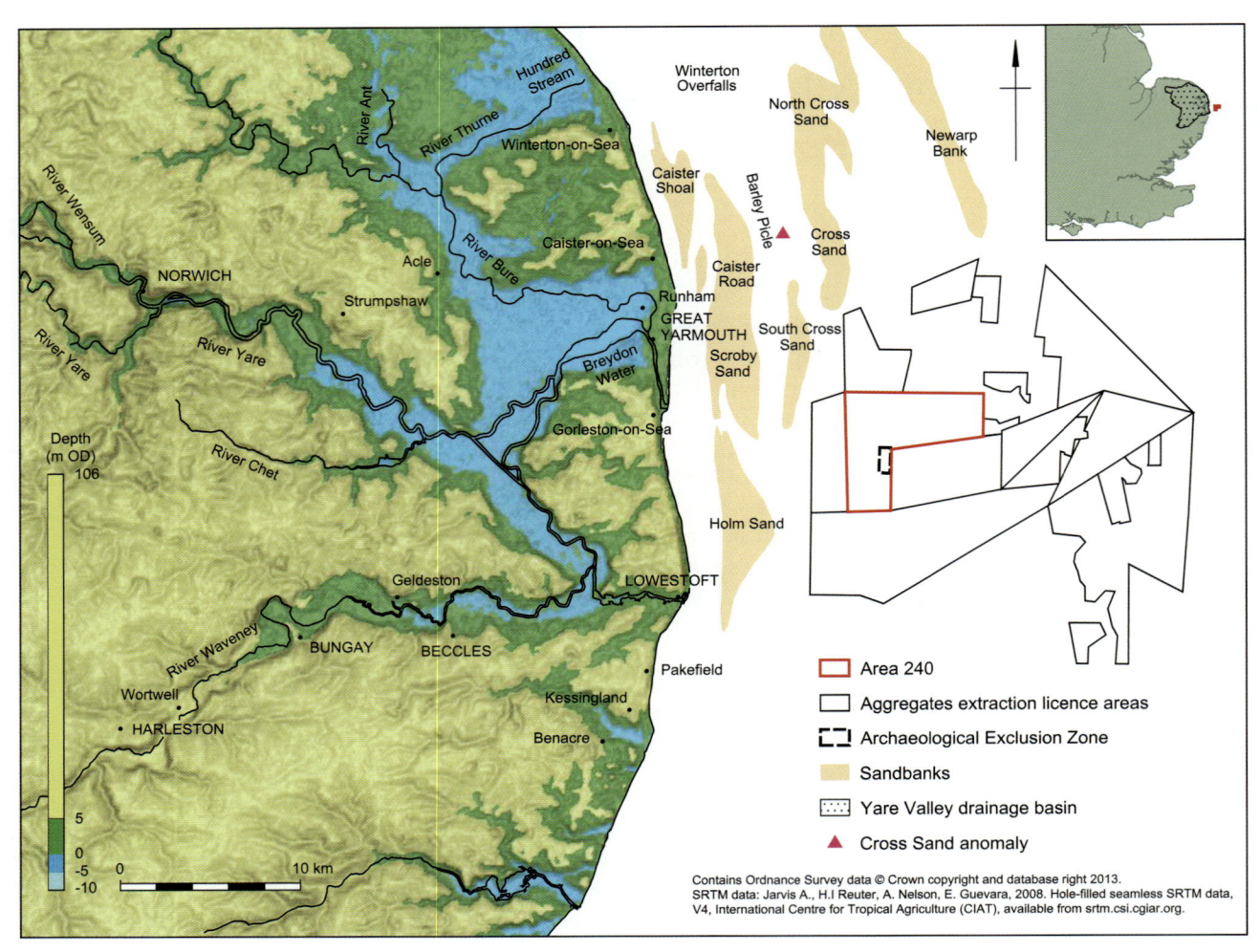

Figure 4.1 Yare Valley location map with place names referenced in the text. Inset: Yare Valley drainage basin

reported from around Norwich, though most reported finds have little to no primary context or associated typology of any detail therefore limiting archaeological assessments.

The following prehistoric characterisation of Area 240 aims to provide stratigraphic and chronological context for the recovered archaeological material. The characterisation is based principally on the results of the palaeogeographic reconstruction of Area 240. However, the identified stratigraphy and inferred depositional environments form part of the larger Palaeo-Yare Valley system and are discussed in this context (Fig. 4.1). The characterisation also provides a regional overview of known environmental conditions in eastern Britain. Patterns of likely hominin presence are briefly discussed, although a comprehensive overview of the British Palaeolithic and Mesolithic is outside the scope of this volume (for examples see, Pettitt and White (2012) for the Palaeolithic and Tolan-Smith (2008) for the Mesolithic).

In order to address the spatial scales involved in the characterisation it is necessary to define terms.

'Area' refers to Area 240 and within the Area there are two zones: Archaeological Recovery Zone (ARZ) from which the original archaeological was dredged in 2007 and 2008 and Archaeological Exclusion Zone (AEZ) put in place by the licensee. As Area 240 is situated in the East Coast aggregate block, reference is made to individual licence areas where appropriate (Fig. 4.1). The present-day Yare Valley system has a drainage basin covering a large part of Norfolk and Suffolk. However, the focus of this characterisation is the area of post-Anglian Pleistocene and early Holocene formations and terrace deposits which are generally constrained by the 5 m OD contour and is referred to as the Palaeo-Yare Valley or catchment area. The Palaeo-Yare Valley extends 30 km west of the present-day coastline and extends approximately 35 km to the east; although the Palaeo-Yare Valley is a single system it is difficult not to refer to it in its present-day state. As such 'onshore' refers to the present-day terrestrial section of the Palaeo-Yare, 'nearshore' is used to define the section between the coastline and the aggregate block; the block itself and open sea are referred to as 'offshore'. The general term 'now-submerged' is used to define both nearshore and offshore environments. The Palaeo-Yare is also set within the wider regional context which includes eastern Britain, the southern North Sea and western continental Europe.

Chronologically, the last one million years is discussed with a focus on the development of the Palaeo-Yare from the late Anglian to inundation of the now-submerged section of the valley some 7000 years ago. The characterisation is based on broad timescales referencing MIS and appropriate sub-stages. This relaxation of time constraints and discussion of long time periods (glacial/interglacial) periods is required to provide an overview of the general trends of marine regression (sea-level fall and exposure of the continental shelf) and marine transgression (sea-level rise and inundation of the coast) throughout the time-scale under consideration. Where ages, from optically stimulated luminescence (OSL) techniques or radiocarbon dating are known, laboratory details are provided; although these tend to be both spatially and chronologically localised, and the OSL data do not necessarily yield more precise ages for Middle Pleistocene sediments.

The Present-day Setting of the Palaeo-Yare

The present-day remnants of the Palaeo-Yare Valley system includes the Rivers Yare, Wensum and Waveney which are fed by a large drainage basin (Fig. 4.1). The lower reaches of the River Yare and Waveney flow into Breydon Water, the remnants of an outer estuary, and flows east from Breydon Water and then south into the North Sea (Fig. 4.1). To the north and south of Great Yarmouth the region is shaped by its rivers and the Norfolk Broads (remnants of peat removal between the 12th and 14th century, since flooded). The coastline to the north comprises a cliff that is generally less than 10 m high and erodes at an average rate of 0.9 m per year (Clayton 1989). To the south of Great Yarmouth the eroding coastline (Pl. 4.1) is cut by the River Waveney at Lowestoft, a man-made river cut from the Waveney 5 km inland to the coast in 1827 to increase trade to the coast and to supplement the harbour at Great Yarmouth.

The geology of the Palaeo-Yare Valley predominantly comprises two sediment formations: Yare Valley Formation and the overlying Breydon Formation. The Yare Valley Formation (Arthurton *et al.* 1994) is observed throughout the river valleys to the north of Kessingland as far as the River Ant to the north of Great Yarmouth. To the south, the Yare Valley Formation is known to extend into the Waveney Valley, however lack of borehole evidence means that the full extent is unknown (Moorlock *et al.* 2000). The

Plate 4.1 Eroding cliff at Pakefield, Suffolk

Yare Valley Formation predominantly overlies pre-Anglian crag deposits and gravelly sediments assigned to the Yare Valley Formation have been recorded in valleys to the east of Newtown, Great Yarmouth, resting on Crag. The thickness of Yare Valley Formation is up to 11 m near Great Yarmouth and comprises fine to coarse gravel with variable amounts of fine- to coarse-grained sand. The gravel, mostly flint and silty gravel, is observed in some cores. Formally, the unit is defined according to a borehole (no. 8) situated at Runham, adjacent to the banks of the River Yare outflow from Breydon Water. The Formation is 5.2 m thick with a maximum depth of 24 m below OD and comprises grey, silty, fine to coarse gravel passing in the uppermost metre to grey-brown gravelly medium grained sand (Arthurton *et al.* 1994).

There is no definitive age for the deposition of the Yare Valley Formation. However, characteristics of the early deposits indicate a possible late Anglian age, MIS 12 (Arthurton *et al.* 1994) with upper deposits suggested as Devensian, MIS 5–2 (Coxon 1979) or late Devensian, MIS 2 (Cox *et al.* 1989). Arthurton *et al.* (1994) supposes at least some of the deposits to be late Devensian/early Holocene age and that at least the basal deposits of the unit may have been deposited during the late Anglian.

There are few terraces associated with the Wensum and Yare Rivers and the onshore sedimentary sequence indicates little significant uplift in eastern Norfolk (Westaway 2009). Onshore boreholes indicate the presence of terrace deposits in the upper reaches of the rivers which form flat or gently sloping features, generally 1 to 3 metres above the later deposited alluvium of the Breydon Formation. Sub-angular and angular flints with some quartz or quartzite pebbles are found on the surface of the terraces. The age of these deposits is uncertain though

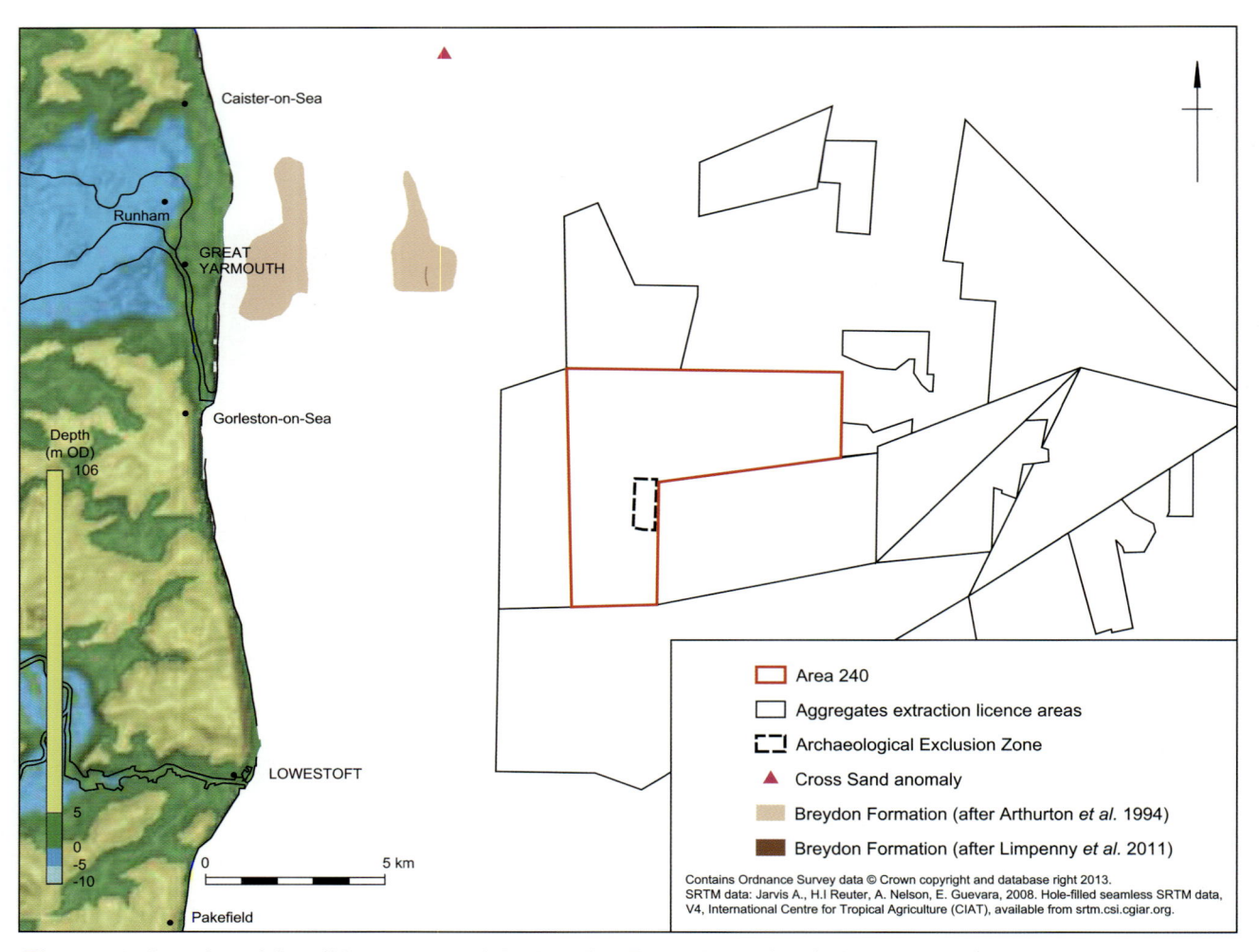

Figure 4.2 Location of the offshore extents of the Breydon Formation (after Arthurton et al. 1994; Limpenny et al. 2011) and the location of the Cross Sand anomaly, the remains of probable chalk raft

Cox et al. (1989) suggested a late Hoxnian (MIS 11 c. 400 ka) age. However, given the ages of the terraces in the River Waveney and the likely cold depositional environment it is possible that these terraces relate to MIS 10 (374–337 ka) or 8 (300–243 ka).

In the Waveney Valley three terrace sets are identified between Wortwell and Geldeston (Moorlock et al. 2000). The terraces have mainly been mapped in the western areas and are absent further downstream. The third (and highest) is the Homersfield Terrace which forms a distinct irregular topographic bench typically 6 m above the present floodplain. Environmental evidence suggests deposition in a cold post-Anglian period (Moorlock et al. 2000) and the height of the terrace suggests a MIS 8 age (Westaway 2009).

The sands and gravels of the Broome Terrace (terrace 2) are situated approximately 4 m above the present-day floodplain and are thought to have been deposited in cold climate during the Saalian, based on reworked Hoxnian interglacial pollen and that the terrace is stratigraphically higher than nearby Ipswichian deposits at Wortwell. This tentatively implies deposition in MIS 6 (191–130 ka) (Coxon 1993). The youngest terrace deposits of the first 'floodplain' terrace are documented as late Devensian in age (MIS 2, 29–11.7 ka). Sand and gravel extraction working within the Waveney Valley indicate that the river terrace deposits contain large flint nodules (0.3 m) clearly not transported far from the chalk source (Moorlock et al. 2000). The gravel content of the terraces varies between 29 and 59% (average 41%) and comprises mainly flint, quartz, quartzite and rare chalk (Moorlock et al. 2000). The Homersfield and Broome terraces converge downstream indicating a greater degree of uplift and terrace development in the west and a tapering to the east (Westaway 2009).

Throughout the Palaeo-Yare Valley system these underlying floodplain sand and gravel deposits are overlain by Holocene post-glacial deposits belonging to the Breydon Formation deposited as the sea levels began to rise after the last transgression (Arthurton et al. 1994). The Breydon Formation is dominated by

Table 4.1 Sediment units identified in Area 240 with onshore and offshore equivalent formations

Area 240 Unit	Proposed age of unit	Environment	Offshore Formation	Onshore Formation
8	Holocene (post-transgression; MIS 1)	Marine	Holocene seabed sediments	
7	Early Holocene (MIS 1)	Fluvial/inter-tidal/early transgression		Lower Breydon Formation (including the Basal Peat Formation)
6	Possibly mid-Devensian (MIS 3)	Alluvium		Possible Yare Valley Formation ('First' terrace)
5	Unknown, possibly contemporary with Unit 6 (MIS 3) or early Devensian (MIS 5c–3)	Possibly represents an estuarine or near coastal depositional environment		
4	Early Devensian (MIS 5c)		Brown Bank Formation	Yare Valley Formation
3b	Middle Saalian (MIS 8/7)	Glaciofluvial		Yare Valley Formation (possible Homersfield terrace)
3a	Possibly late Anglian (MIS 12) or early Saalian (MIS 10)	Glaciofluvial		Yare Valley Formation
2a/b	Cromerian complex (MIS 26–16)	Delta top/shallow marine	Yarmouth Roads Formation	Cromer Forest-bed Formation (partially equivalent)
1	Pliocene/Early Pleistocene	Shallow marine	Westkapelle Ground Formation	Upper Crag Group (partially equivalent)

silt and clay deposits and three peat layers: the Basal, Middle and Upper Peat. The Basal Peat Member of the Breydon Formation is reported to have formed 6600 to 6240 cal BC (7580±90 BP, HAR–2535) at a depth of around 19 m below OD and is up to 2 m thick (Arthurton et al. 1994). Based on seismic data of the near coastal area, the Breydon Formation is thought to be preserved offshore in two distinct areas off Great Yarmouth, approximately 6 km east of the present-day coastline, as illustrated in Figure 4.2 (Arthurton et al. 1994). The Middle Peat Member formed between 4700 and 2200 BP. The Upper Peat Member developed around 1750 BP (Boomer and Godwin 1993; Arthurton et al. 1994).

Nearshore, the seabed is dominated by a series of large sandbanks that were formed since the last marine transgression. These banks, such as Cross Sand, Scroby Sand and Caister Shoal (Fig. 4.2), are up to 25 m high and are composed of material generated from the eroding coastline to the north and erosion of localised seabed sediments (D'Olier 2002). It is possible that the banks originally formed as banner, or headland banks, which are long banks of sand that lie with one end almost connected to the coast at headlands.

Beyond the banks lies a gently sloping seabed deepening to approximately 40 m below OD before deepening to 50 m below OD at an apparent break of slope 40 km from the coast. The seabed morphology continues to change with mobile sandwaves affected by seabed currents and new bedload material transported into the nearshore area from the north.

The group of aggregate licence areas are situated between 8.5 and 30 km from the coast. The area has been dredged since the 1970s and the dredging has changed the seabed morphology in this area. The seabed in the licence aggregate block is generally between 20 and 35 m below OD, excluding effects of mobile sandwaves. Area 240 is situated in the west of the aggregate block in water depths of between -16.7 and -33.5 m CD (18.2 and 35.0 m below OD). The area is dominated by a series of west to east orientated sandwaves, up to 6 m high.

Area 240 Palaeogeographic Reconstruction

The sub-seabed sediments observed in Area 240 are an offshore extension to the Palaeo-Yare river system and, interestingly, provide a clearer pattern of depositional age than the adjacent material found onshore. The assessment of geophysical and geotechnical data, combined with palaeo-environmental and dating techniques have allowed an in-depth assessment of the sediment units (in more detail than available onshore) and provide further details on the development of the Palaeo-Yare system from its incision during the late Anglian (MIS 12, 424 ka) through to the last transgression (MIS 1).

The palaeogeographical reconstruction of Area 240 reveals a complex history of deposition and erosion. Eight sediment units have been identified, dating from the Late Pliocene/Early Pleistocene (Unit 1) to marine deposits associated with the last transgression during the Holocene (Unit 8) (Table 4.1). Figure 4.3 illustrates a schematic of the development of the Palaeo-Yare in Area 240 and the lateral extents of the associated units. Figure 4.4 provides an illustrated profile of the late Anglian channel and infill sediments.

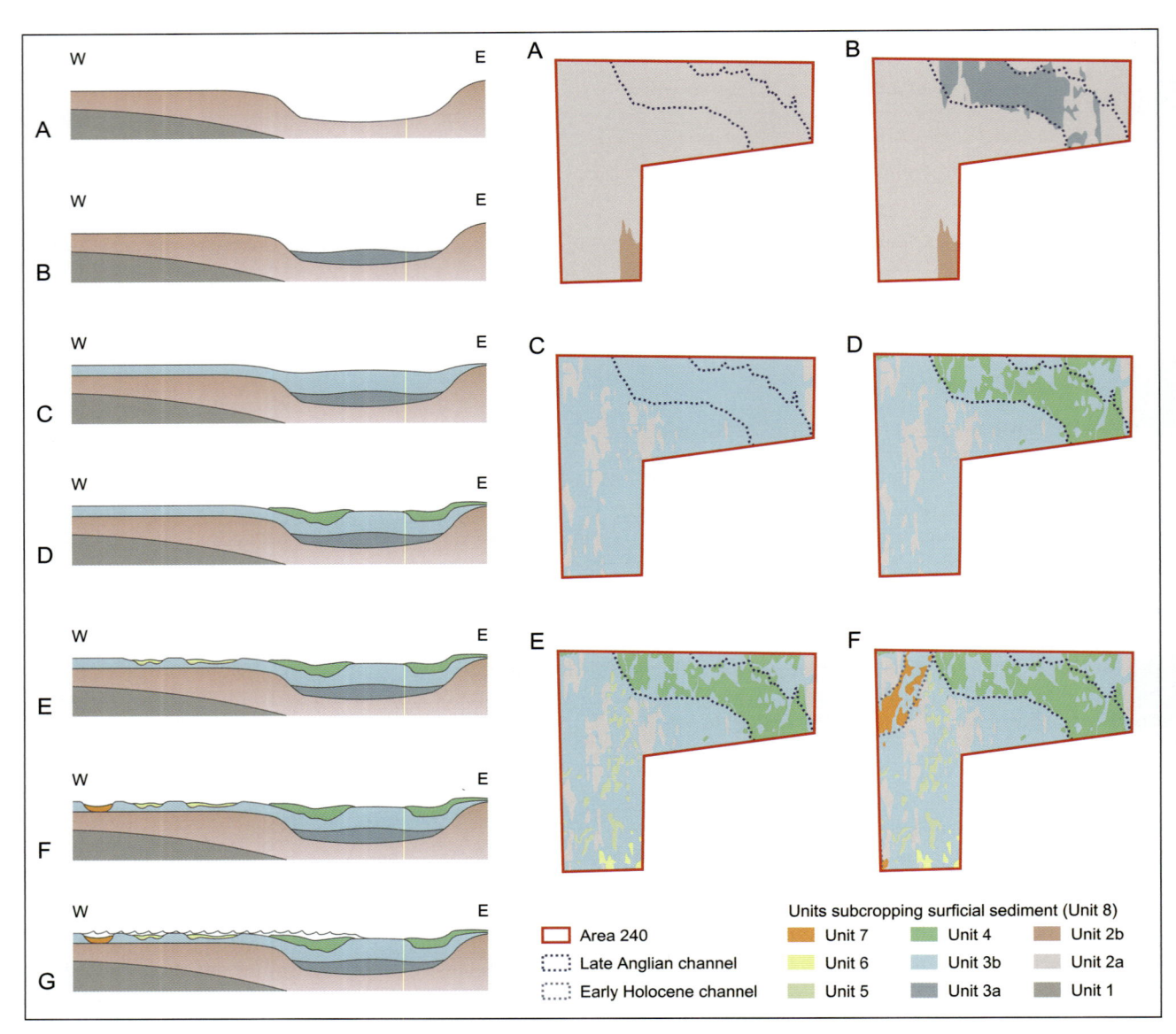

Figure 4.3 Schematic illustrating the development of Area 240 and plan overview. Reference to the development stages are provided in the text

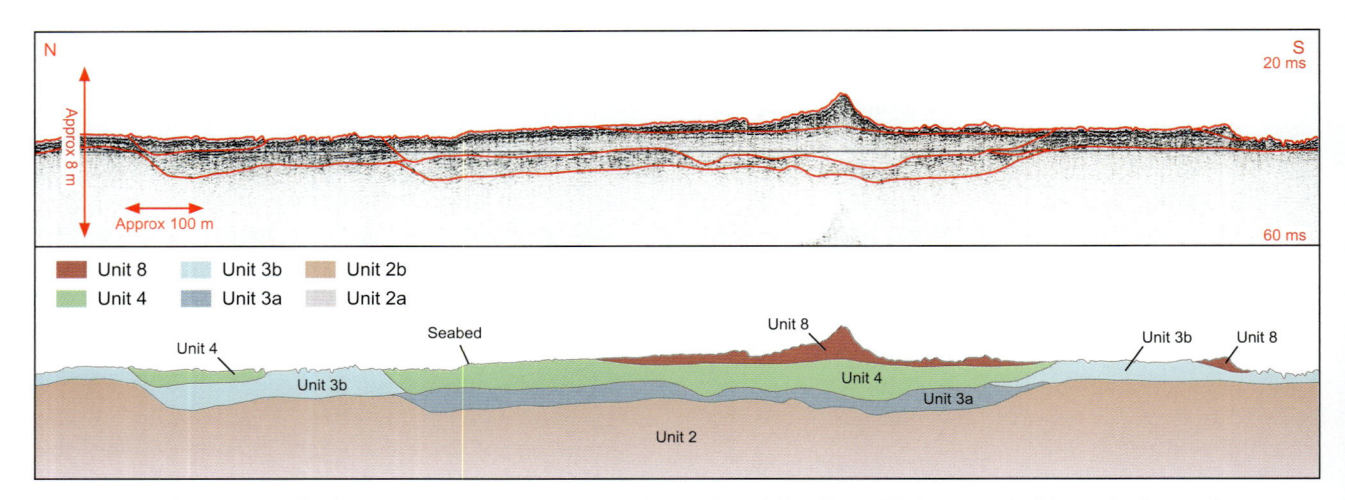

Figure 4.4 Seismic profile illustrating the late Anglian channel and Saalian (Unit 3a and 3b), early Devensian (Unit 4) infill sediments overlain by marine seabed sediments (Unit 8)

Seven stages of development have been identified:

A) Initial development of a late Anglian channel and floodplain base cutting into the underlying pre-Anglian deposits (Unit 2a and 2b);
B) Saalian deposition of basal channel deposits (Unit 3a);
C) Continued channel infill and development of the floodplain deposits (Unit 3b) during the Saalian;
D) Re-activation and infill of the channel (Unit 4) during the early Devensian;
E) Estuarine sediment (Units 5 and 6) deposited during the mid-Devensian across localised areas of the floodplain;
F) Development of early Holocene channel and partial infill (Unit 7);
G) Post-transgression deposition and development of the seabed.

Pre-Yare Valley Palaeogeography

In order to provide context to the development of the Palaeo-Yare Valley system it is necessary to go further back in time and assess the landscape into which the Palaeo-Yare was initially cut.

The Ur-Frisia Delta Plain: Pre-Anglian (MIS 13 upwards; >478 ka)

During the majority of the Pliocene (c. 5.3 to 2.6 Ma) Britain was surrounded by warm temperate seas (Funnell 1995) and throughout this time Area 240 was situated in a shallow marine environment with the coastline approximately 60 km to the west. Then, commencing around 2.6 Ma, the sea level fell in association with the northern hemisphere glaciation.

A series of marine transgressions and regressions throughout the Early Pleistocene resulted in the deposition of the Crag Group (Red Crag, Norwich Crag and Wroxham Crag) in a shallow shelf and intertidal environment comprising sands, silts and clays with occasional gravels (Moorlock et al. 2000). Remnants of these deposits are observed onshore and the western limit of the Group marks the westernmost limit of the coastline at 2.6 Ma. From around 2.6 Ma to 1.7 Ma there was an overall trend of northward regression (Cameron et al. 1992) and the development of a northward progressing delta. During the overall regressive trend, there were fluctuations and the sea transgressed and regressed several times (Funnell 1995). By 1.7 Ma, progressive northward deltaic progradation continued and the Ur-Frisia delta top linked Britain to mainland Europe. This excluded all marine influence from the southern North Sea Basin between approximately 1.7 Ma and 500 ka. The Ur-Frisia delta was fed by the Bytham, Thames, Rhine, Meuse and northern German rivers to the coastline situated north of present-day Aberdeen at 57°N (Funnell 1995).

As part of the Ur-Frisia delta plain the Yarmouth Roads Formation was deposited between 2.6 Ma and 478 ka; the later deposits of the formation equate, in part, to the Cromer Forest-bed Formation which have been associated with archaeological material at Pakefield (Parfitt et al. 2005). The Yarmouth Roads Formation is known to comprise a complex delta top sequence consisting of sands with pebbles (including chalk), abundant plant debris and peat clasts (Cameron et al. 1992).

During this period, Britain was a peninsula of Northwest Europe, even during periods of high sea level. During the Early and early Middle Pleistocene, tectonic activity in the form of differential subsidence and uplift occurred. Based on the elevations of marine sediments, uplift of approximately 30 m occurred around Norwich but very little change in elevation around West Runton, Happisburgh or Lowestoft, indicating that (co-incidentally) the present-day coast represents a hingeline with uplift to the west and subsidence to the east (Rose 2009).

On the landward side of the delta top there was a hiatus in major sediment influx accompanied by widespread soil formation throughout south-east East Anglia. Two major rivers flowed through the landscape during this period. The Bytham River flowed eastwards from the Midlands through East Anglia and into the North Sea and the ancestral form of the River Thames flowed from its source in the Cotswolds along its present-day course to Reading and then flowed north-east to the Norfolk coast (Fig. 4.5). At Happisburgh Site 3 a series of channel sediments and associated overbank alluvium belonging to the ancestral River Thames has been mapped and are thought to be either MIS 25 (c. 970 ka) or MIS 21 (c. 850 ka) (Parfitt et al. 2010).

By c. 700 ka the Bytham had changed its course and flowed through Pakefield into the North Sea and the Thames had migrated further south (Parfitt et al. 2005, Fig. 4.5). Early hominin activity at Pakefield is indicated by artefacts recovered from floodplain deposits dated to c. 700 ka (Parfitt et al. 2005, 2010; Lee et al. 2006). The assemblage at Pakefield comprises 32 worked flints, including a simple flaked core, a crudely retouched flake and a quantity of waste flakes (Parfitt et al. 2005). The artefacts were all found in clear stratigraphical contexts relating primarily to the interglacial infill of a channel. The infill comprised extensive deposits of organic muds and clays (Cromerian Complex) deposited within incised late Pliocene and Early Pleistocene marine Crag deposits. The organic sediments of the Formation have also yielded a wealth of faunal

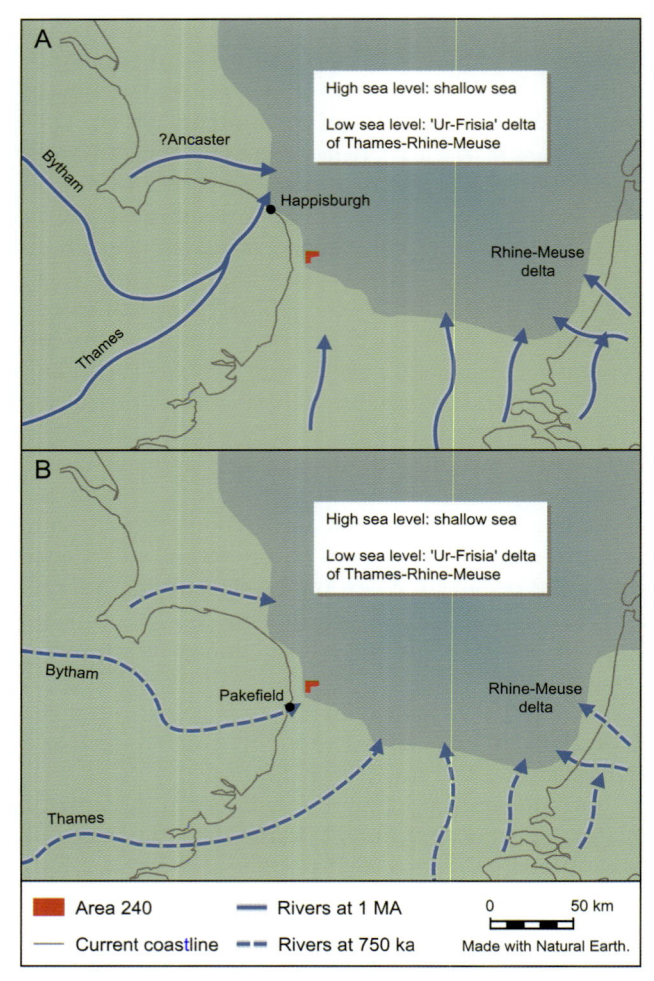

Figure 4.5 Generalised palaeogeography of the Middle Pleistocene illustrating the coastline and major rivers at approximately A) 1 MA and B) 750 ka (after Cameron et al. 1992; Parfitt et al. 2010; Hijma et al. 2012)

evidence consisting of elephants, deer and other large mammals (Wymer 1999). The Cromer Forest-bed Formation is overlain by the terraced river deposits of the Bytham sands and gravels (Rose et al. 2002). It is postulated that three terrace deposits are dated to end-MIS 14 (incision), the temperate climate in MIS 13 and the cold climate in the early MIS 12 cooling transition, immediately prior to the earliest Anglian glaciation (Westaway 2009). However, there is considerable and on-going debate regarding the chronology of this regional stratigraphy and the implications for the Lower and Middle Palaeolithic archaeological record of East Anglia (Hosfield 2011; Pettitt and White 2012).

In Area 240, two sediment deposits relating to the pre-Anglian are observed. Unit 1 is interpreted as the Westkapelle Ground Formation which is partially equivalent to the younger Crag deposits. The Formation was deposited in a pro delta-environment and generally comprises silty clays with partings of sand passing upwards into predominantly mud-free sands (Cameron et al. 1992). Unit 1 is the deepest unit and is observed across Area 240. The top of the unit deepens to the east and westerly dipping sediments are observed within the unit.

Unit 2 is interpreted as the Yarmouth Roads Formation and is divided into two sub-units: Unit 2a and 2b. Throughout the majority of Area 240, Unit 2a is observed overlying Unit 1 and the sediment unit thickens to 18 m in the east. The composition of the unit is variable across the area, predominantly comprising fine- to coarse-grained sands with localised thin laminae of clays and silts which may have been caused by tidal rhythmitic deposition.

Pollen, recovered from this unit within vibrocore VC2c (Fig. 4.6), were dominated by *Pinus sylvestris* (pine), Cyperaceae (Sedge family) and Poaceae (grasses), but the pollen counts are relatively low. A proportionally high number of pre-Quaternary spores and indeterminable grains were observed in comparison to identifiable pollen grains present, indicating that the amount of reworked sediment is probably significant. Where present, the grains of Poaceae and Chenopodiaceae were much better preserved, with a possible implication that they were local in origin and thus representative of the local vegetation. This would have probably been an open brackish/estuarine environment, a suggestion supported by the presence of *Plantago maritima* (sea plantain). The single occurrence of *Dryas octopetala* (mountain avens) may also imply that this was a cold environment.

Foraminifera indicative of shallow marine and outer estuarine environments were recovered from VC8c1 and included species of *Elphidium* and *Elphidiella* (*Elphidium arcticum* and *Elphidiella hannai*) which are usually associated with pre-Anglian cold stages of the Early Pleistocene (Funnell 1989). Molluscs including Venerupidae (carpet shells) also indicate a shallow marine depositional environment.

Optically stimulated luminescence (OSL) dating of this unit in VC3b at 33.57 to 33.67 m below OD returned a date of 735±134 ka (MIS 19; GL 10040) which is considered to be a minimum age estimate indicating a probable Cromerian Complex (MIS 16–22, 621–866 ka) or possibly early Bavelian Complex age (MIS 22–26, 866–959 ka). Due to a relatively low environmental-dose rate received by sample GL 10040 from the surrounding sediments, saturation of quartz dosimeters has not occurred, extending the datable range of this material (Toms 2011). This date is comparable to the age of Yarmouth Roads Formation in the Dutch sector at similar latitudes (Zagwijn 1985).

In the south-west corner of Area 240 there is a distinct change in structure of the unit (Unit 2b) where up to 11 m of southerly-dipping sediments

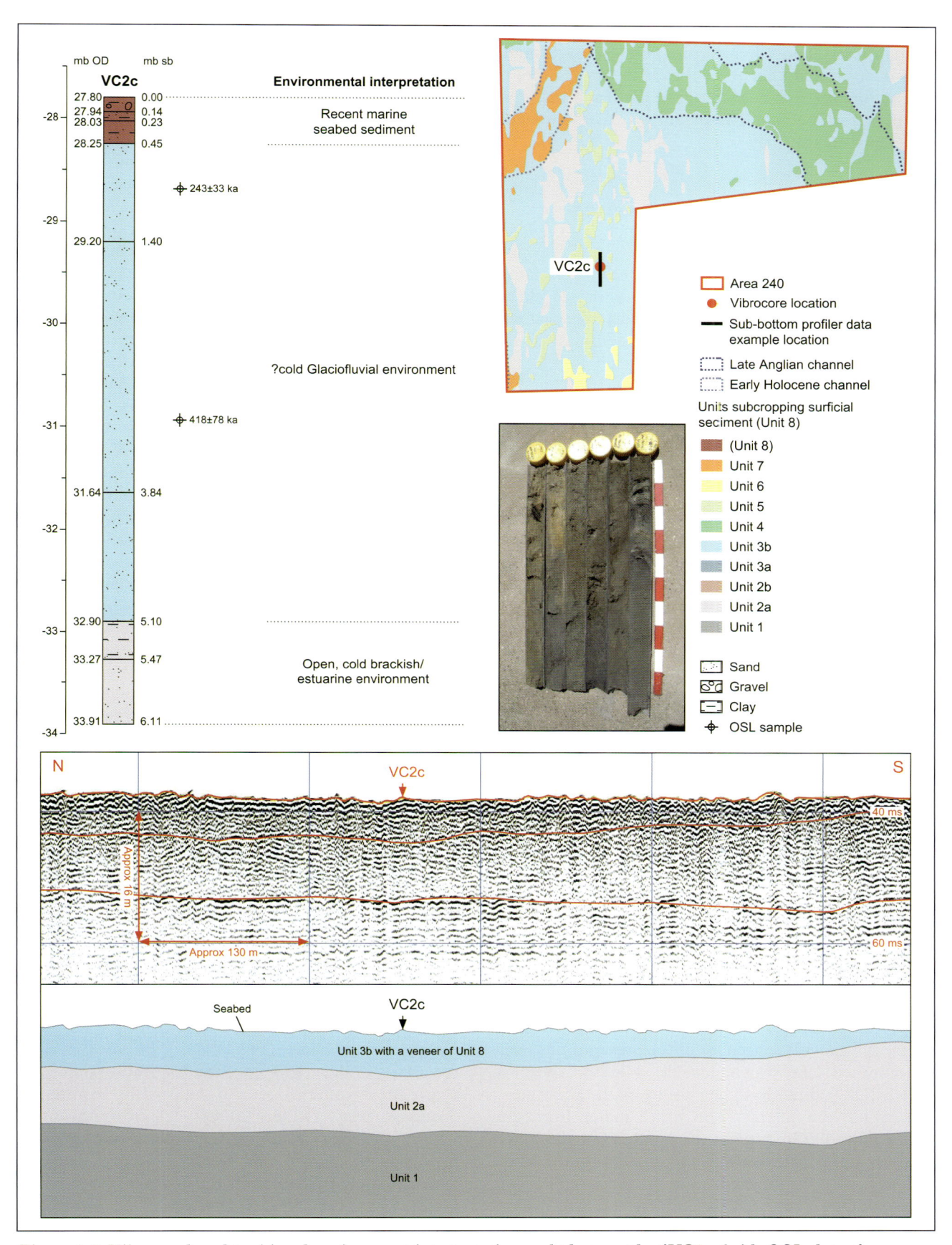

Figure 4.6 Vibrocore log, depositional environment interpretation and photograph of VC2c (with OSL dates from VC3b) with a seismic section illustrating the targeted floodplain sediments

were observed. However, both are considered contemporary and the structure change is attributed to a depositional change of the delta top sediments. The sediments generally comprise silty sand with very frequent thin beds and laminae of firm to stiff clay and peaty organic clay.

Extensive Remodelling of the Landscape: Anglian (MIS 12; 478–424 ka)

The Anglian represents the most extensive glaciation of the British Middle Pleistocene, with ice sheets reaching down as far as the north Cornish coast and the Thames Valley, and extending north-east to the Continent (Huuse and Lykke-Andersen 2000). The glaciation had a marked effect on the landscape and radically altered drainage patterns and relief (Lee et al. 2011). Extensive remodelling of the landscape took place, with old river courses such as the Bytham River destroyed or buried by till deposits. Continental-wide change took place with the large European rivers, such as the Rhine, diverted south and the Baltic/Eridanos River, which previously flowed from western Russia and Scandinavia to the North Sea, replaced by the Baltic Sea (Rose 2009). Sediment supply to the southern North Sea delta was greatly reduced with sediment redistribution rather than accumulation. The trapping of water within the extensive Anglian ice sheets resulted in a fall in sea level thought to be the lowest recorded around the British Isles and estimated at 130 m below the present level.

The southern limit of the Anglian glacier has been mapped onshore based on the presence of end moraines and incised tunnel valleys. Offshore, the southerly limit is associated with morphologically similar buried, subglacial tunnel valleys, glaciotectonic deformation structures and subglacial 'till tongues' (Graham et al. 2011). The consequence of this extent was the re-directing of the European drainage network south of the ice margin. Recent work has indicated that the limit of the ice may in fact be further south than originally thought (EMU Ltd 2009; Dix and Sturt 2011).

There is much debate about the landscape within the southern North Sea region during the Anglian. There is a general consensus of the presence of a large ice-dammed lake which developed in the southern North Sea – directly to the south of the ice-front – into which the Thames and other major European rivers flowed (Gibbard et al. 1988; Gibbard 2001; Murton and Murton 2012). To the south of the lake a spillway was established over a topographic low between Dover and Calais and created new valleys (Gibbard et al. 1988; Gupta et al. 2007, Hijma et al. 2012). However, the sedimentological evidence for the presence of such a large lake is lacking (Laban and van der Meer 2011). In either scenario, Area 240 would have been covered by ice during the period of the glacial maximum and the pre-Anglian landscape in the immediate area would have been altered.

The immediate post-Anglian landscape in the region to the north of the maximum ice limits would have been dominated by till plains, with newly exposed glacial landforms such as kettle holes and over-deepened glacial valleys, the latter often forming the basis for new drainage systems (Wymer 1999; Pettitt and White 2012). Onshore till deposits belonging to the expansive Corton Till Member of the Happisburgh Formation and overlain by the Lowestoft Till Formation are observed in East Anglia and probably extended offshore although the extent of these till deposits is difficult to predict. Although the age of the Lowestoft Till Formation is accepted as MIS 12 there is a some debate over the age of the underlying Happisburgh Formation which may have been deposited as early as MIS 16 (Rose et al. 2001; Lee et al. 2004; Rose 2009) indicating multiple phases of remodelling of the region, although there are those who disagree and argue for a MIS 12 age largely based on pollen biostratigraphy (Preece and Parfitt 2008).

There is no doubt that the Anglian glaciation had a major impact on the existing Cromerian Complex deposits and it is probable that younger Cromerian Complex deposits (the younger Yarmouth Roads Formation) were eroded during this time and that tills would have been deposited, at least in part, in the offshore coastal region. There is no evidence of any till deposits surviving in the offshore coastal region and it is likely that any such material was eroded during subsequent periods of high sea level, in particular the Hoxnian (MIS 11) and Ipswichian (MIS 5e) interglacial stages. Based on current rates of cliffline erosion it has been surmised that around 10,000 years ago the 'cliffline' was situated approximately 8 km east of the present-day coastline and has since been eroded away under marine conditions (D'Olier 2002). It is likely that this cliffline was an extension of the coastline today and comprised glacial till. D'Olier (2002) suggested that an upland area comprising till overlying crag was present between Winterton-on-Sea and Benacre possibly extending north-eastwards as far as the Newarp Banks, Winterton Overfalls and North Cross Sand (Fig. 4.1). The erosion and reworking of these tills since the last transgression are a possible source for the nearshore sandbanks observed today (D'Olier 2002).

This theory is tentatively supported by the presence of the 'Cross Sand anomaly' situated approximately 6.8 km from the coast to the north of Great Yarmouth, 9 km north-west of Area 240 (Fig. 4.2). The feature is approximately 165 m long, 30 m wide and 13 m high. The true nature of the feature is unknown; however one theory is that the

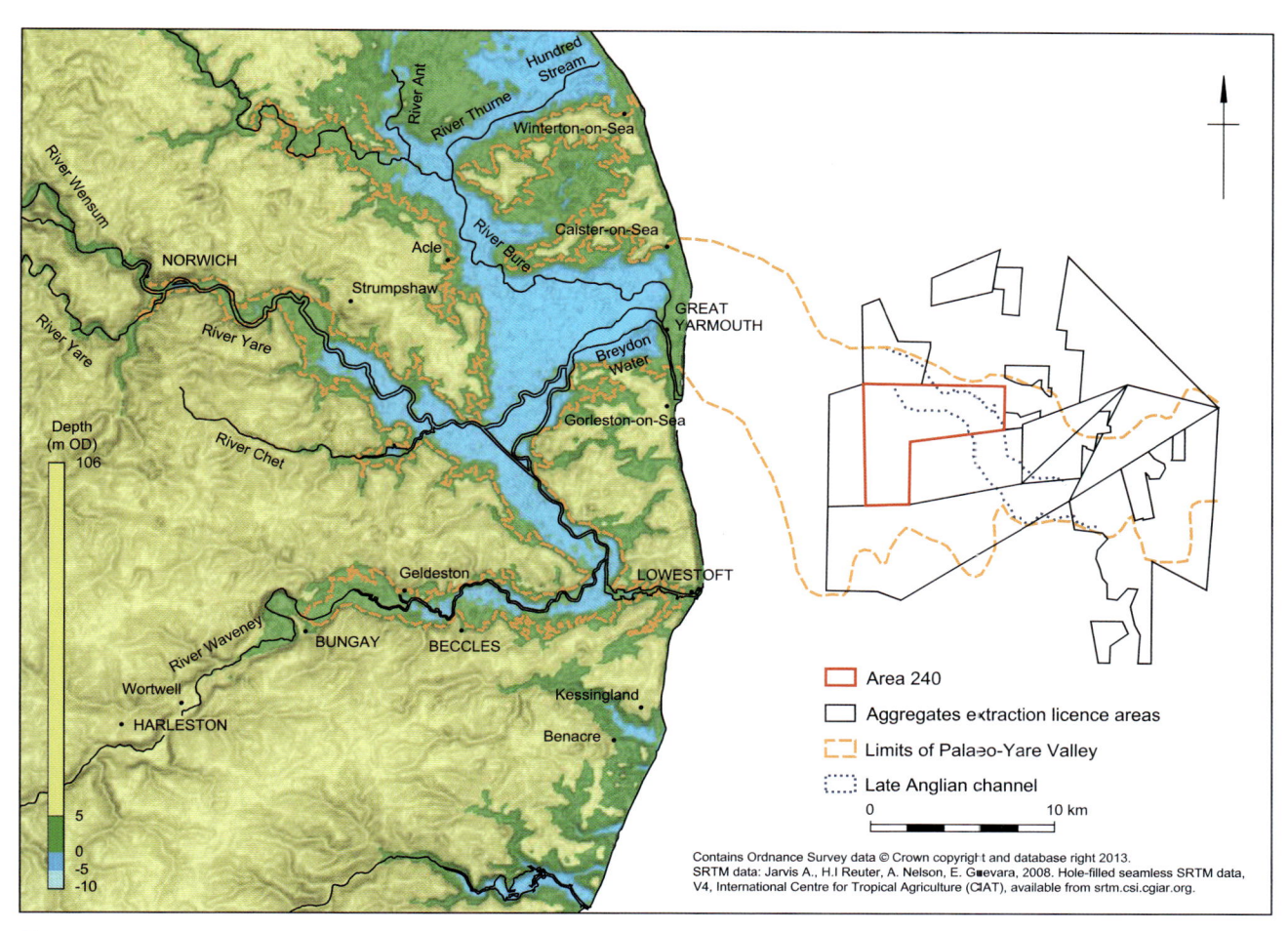

Figure 4.7 Limits of the Palaeo-Yare floodplain and relation to the late Anglian channel

feature is a chalk raft deposited during the Anglian (Limpenny et al. 2011). Chalk rafts can be seen on the north Norfolk coast and are defined as dislocated slabs of bedrock and/or unconsolidated sedimentary strata that have been transported from their original position by glacial action (Burke et al. 2009). On the north-east coast of Norfolk the chalk rafts are exposed within the Middle Pleistocene till deposits that are exposed in the cliffs capped by periglacial sand and gravel. It is possible that the Cross Sand chalk raft formed part of the cliffline at one point and has subsequently been eroded away leaving the feature exposed. This would indicate that the till deposits were deposited at least 6.8 km east from the present-day coast.

Early Development of the Palaeo-Yare

A New Drainage Pattern: Late Anglian (MIS 12; c. 434 ka)

When the Anglian ice started to melt after the glacial maximum (c. 434 ka) the present-day drainage pattern developed on a predominantly till-covered landscape. Some rivers, such as the upper reaches of the Waveney, broadly re-occupied remnants of their pre-glacial channels, while other utilised the new topography of meltwater channels (Moorlock et al. 2000; Gibbard and Clark 2011). However, Rose (2009) suggests that the course of the pre-glacial river systems did not influence the post-glaciation river development and that the parallelism of the River Waveney (amongst others) and the course of the pre-glacial Bytham River is simply co-incidental where subglacial erosion or meltwater incision has cut valleys across or along the preglacial valley and this subglacial valley has subsequently been used by subaerial drainage across the post-glacial landscape. Co-incidental or not, from Diss eastwards, the River Waveney follows the lowermost 70 km of the Bytham River to the present-day coast (Westaway 2009). The development of the Palaeo-Yare was initiated during this time and covered much the same area as is observed today with the Rivers Wensum, Yare and the Waveney flowing into a wide floodplain (now Breydon Water) and continuing eastwards to the lower reaches of the Yare Valley which are now submerged. The limits of the floodplain are thought to extend east through the nearshore area although there is limited

Figure 4.8 Model of the base of late Anglian channel feature looking north-west illustrating the broad, shallow nature of the channel (vertical exaggeration: x15)

evidence. The northern and southern limits of the valley area are obscured by the presence of large overlying sandbanks. Regional geophysical data cannot confirm the presence of the Palaeo-Yare Valley deposits, partly due to sub-seabed penetration issues and partly due to the nature of the sediments and the similarity with the underlying basement sediments (probable Unit 1). However, the base of the sandbanks has not eroded to a level below the projected valley floor and as such, valley sediments may be preserved (Wessex Archaeology 2013b).

Within the aggregate block the limit of the floodplain is observed to approximately 28 km from the present-day coastline and is up to 14 km wide in places (Fig. 4.7). Within the limits of the floodplain a channel is observed with a 'steeper' inside and 'gentle' outside sloping edges (Fig. 4.8). The channel feature is observed orientated west-east in Area 240. It continues south through Area 228 and then east in Area 251 (Fig. 4.7). Beyond there, evidence for the broad floodplain, but not the channel, can be found. This is either due to a levelling out of the channel edges (so they are not obvious in the geophysical data) or the removal of the edges due to subsequent reworking and erosion.

The slope of the floodplain floor is relatively gentle. Based on (admittedly few) onshore boreholes and offshore vibrocores and notwithstanding any uncertainties associated with uplift and subsidence, the base of the floodplain is a gentle slope (approximately 0.4 m km^{-1}) from 8.5 m below OD at the confluence of the Rivers Yare and Wensum to 38.5 m below OD at the easternmost limit of the known floodplain (Wessex Archaeology 2013b). As a point of comparison, the (modern) River Thames has a downstream gradient in its lower reaches of approximately 0.3 m km^{-1}.

To the north of the floodplain (between Winterton-on-Sea and Caister-on-Sea), the upland area comprising till overlying crag, possibly extended north-eastwards as far as the Newarp Banks. Tributaries from this upland area probably supplied the River Yare (Fig. 4.9). It is likely that small waterways such as the Hundred Stream and the Thurne River flowed southwestward toward the River Bure from higher ground that was once situated offshore (D'Olier 2002). D'Olier postulated that parts of the deep channels of Barley Picle and Caister Road, between these banks, were the locations of streams that once ran off southwards from the higher ground into the River Yare. On reviewing the geophysical data that were acquired within the area of Barley Picle in 2006 (Wessex Archaeology 2008c), there is some evidence of a very minor cut and fill of sediments between the sandbanks in the south of the area. However, it is not possible to ascertain whether this is related to a Middle Pleistocene river or associated with post-transgression erosion and re-deposition of sediments.

Tills would have extended to the south of the floodplain with an east-west aligned watershed that extended through the high ground, north of Kessingland. Small streams probably ran north or north-east from this upland into the River Yare. There is a possibility that the Waveney River ran directly to the east rather than joining up to the Yare (at Breydon Water) through the lowland area observed today. If so, it is possible that these lowlands at Lowestoft, and to the south-east of Lowestoft, would have been tributaries to the now-submerged Palaeo-Yare.

A Marine Incursion: Hoxnian (MIS 11; 424–374 ka)

At the end of the Anglian the climate warmed and the Hoxnian interglacial followed. Initially, sea levels rose rapidly and by 400 ka were around 10 m below OD. For much of the Hoxnian, Britain would have been cut off from the Continent. However during the latter parts of the interglacial, as temperatures began to cool and sea levels lowered, Britain is likely to have once again been a peninsula of Northwestern Europe. Typically, the Hoxnian shows a classic succession of vegetation that one would expect from an interglacial: open grassland was followed by pine-birch coniferous forest, then by fully temperate deciduous oak woodland, which gave way to boreal forest as soils degraded and climate deteriorated (Pettit and White 2012).

The record of hominin movements and activity in the Hoxnian interglacial is fragmentary but coastal plains and river valleys were key locations for early hominin activity (Pettit and White 2012). The Lower Palaeolithic sites of the Palaeo-Yare appear to have been particularly rich in implements; excavations at Whitlingham, on the River Yare, 4 km east of Norwich, produced several hundred lithic artefacts (Wymer 1999; Sainty 1927; 1933). Further afield, at Beeches Pit in Suffolk, there is evidence of flint knapping and the use of fire occurring adjacent to an abandoned channel of the Bytham River, dated to approximately 400 ka (Gowlett 2006). At Hoxne, a complex sequence of laminated lacustrine sediments

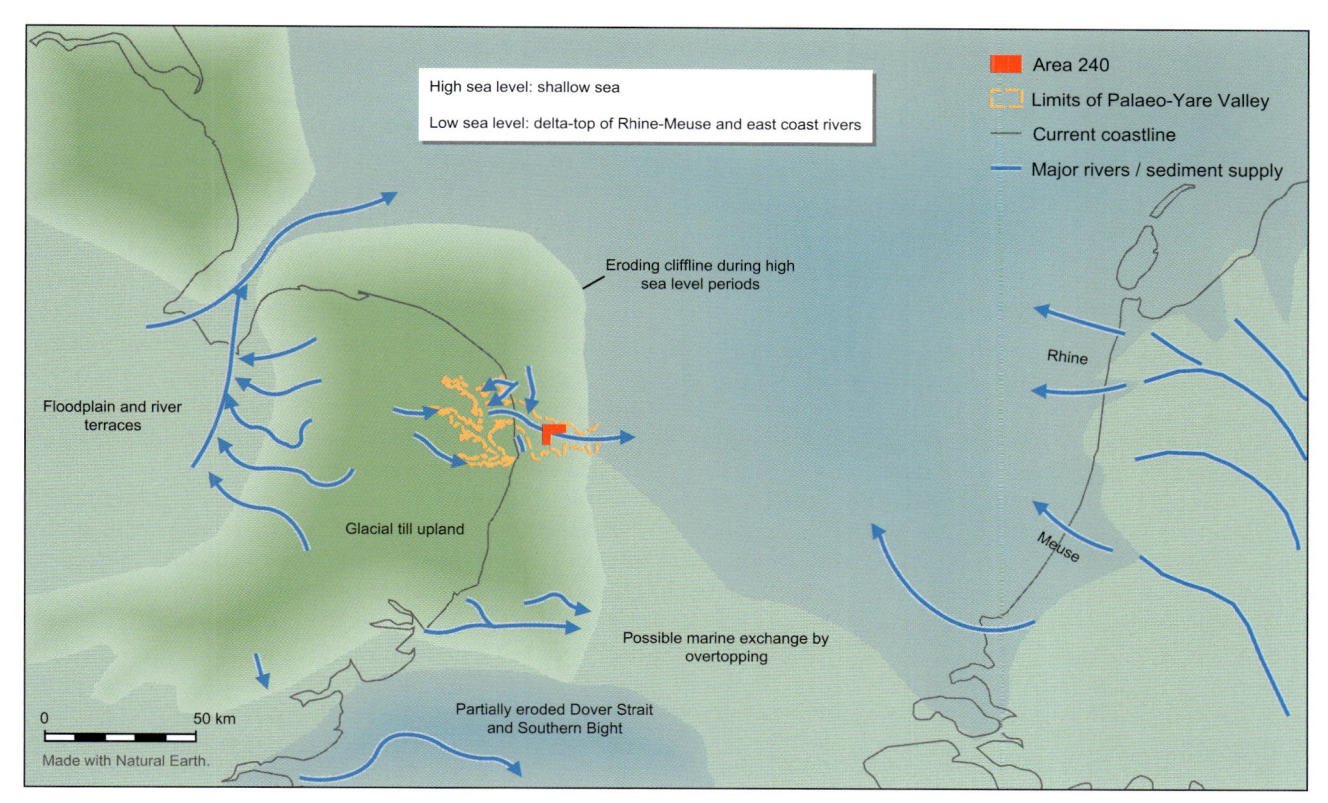

Figure 4.9 Generalised palaeogeography from the late Anglian (MIS 12) to the Saalian (MIS 6) (after Limpenny et al. 2011; Hijma et al. 2012)

was deposited as a lake formed in a kettle hole, which formed as the Anglian ice sheet retreated. Large numbers of hand axes in near mint condition indicated *in situ* deposits, associated with mammalian teeth and bones with indications of deliberate cut-marks (Ashton *et al.* 2008). At Swanscombe, Kent a hominin skull confirms that *Homo heidelbergensis* was active in southern Britain (Wymer 1968). Further evidence, a tibia and incisors, were recovered from Boxgrove, Sussex (Roberts and Parfitt 1999). At Southfleet Road, Ebbsfleet, Kent a complex sequence of fossiliferous Middle Pleistocene (almost certainly dating to MIS 11) sediments were recorded that contained lithic artefacts associated with faunal evidence of elephant butchery (Wenban-Smith *et al.* 2006; Wenban-Smith 2013).

During the Hoxnian the lower reaches of the Palaeo-Yare became increasingly estuarine, and then a shallow marine environment, as the sea level continued to rise. The upland areas of till to the north and south of the floodplain would have formed the coastline and probably would have undergone at least some degree of erosion. The sedimentary record for this period is fragmentary in Norfolk (Arthurton *et al.* 1994). No deposits that can definitely be described as Hoxnian warm stage have been identified; although sands and organic silts, infilling a channel cut in the Lowestoft Till at Caister-on-Sea, seem likely to be of this age and small outliers of Hoxnian sediments were preserved within deposits on top of the till in northern Suffolk (Moorlock *et al.* 2000).

In Area 240, any sediment associated with shallow marine deposition or subsequent regression at the start of the Saalian period is no longer preserved. Finer-grained deposits may have been removed from the area and there may have been some reworking of these sediments with the upper units of any late Anglian deposits during the regression. Any Hoxnian sediments deposited on top of the upland tills in the offshore area would have been removed by subsequent erosion and reworked throughout successive transgressions up to, and including, present-day.

Channel and Floodplain Development During the Saalian (MIS 10–6; 374–130 ka)

Overview

The Saalian period saw major alternating periods of warm and cold with fluctuating sea levels and climatic conditions (Fig. 4.10). There is much debate over the timing of the major Saalian Glaciation (MIS 10, 8 or 6; as discussed in Gibbard and Clark 2011) which is not well represented in the Pleistocene record

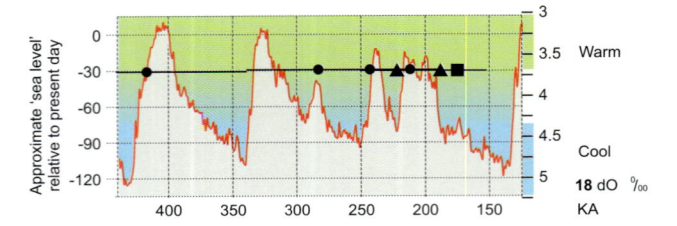

Figure 4.10 MIS curve for the Saalian (MIS 10–6; 374–130 ka) illustrating OSL dating results for this period

compared with the Anglian or Devensian glaciations which have been defined in greater detail. Despite this, it is generally considered that the major glaciation occurred in MIS 6 with two severe cooling periods during MIS 10 and 8. In north Norfolk, stratigraphy of the upper succession of the till deposits of the Sheringham Cliff Formation and Britons Lane Formation are possibly datable to MIS 10 and 6, respectively (Rose 2009). This is, however, disputed and conventional interpretations imply deposition during MIS 12.

Cooling into MIS 10 was a slow process. Oscillations between cold and warm phases occurred before the onset of more extreme cold conditions during MIS 10. Sea level at this time was 100 m lower than present-day. MIS 8 lasted c. 50,000 years, however the climate was less severe than most other Middle Pleistocene glaciations (Tzedakis 2005). Beets et al. (2005) suggest a MIS 8 date for till deposits in the eastern southern North Sea, though there is no evidence to suggest extensive coverage over the southern North Sea. During MIS 6 a substantial ice-lobe advanced down the eastern side of Britain and filled the Fenland Basin where it dammed a series of westward-flowing streams (Fig. 4.11). The result was the formation of shallow glacial lakes that coalesced, culminating in an extensive proglacial lake (Gibbard and Clark 2011). The lake drained westwards to the North Sea via the River Waveney. The Waveney/Yare Valley, in turn, drained into a lake that formed in the southern North Sea. This was comparable, though smaller, to that which was present during the Anglian glaciation (Busschers et al. 2007). This ice-dammed lake would have been surrounded by a tundra and open steppe environment.

During the MIS 9 (Purfleet) interglacial, hominins were active in northern Europe, and in southern Britain many Palaeolithic sites and findspots have been recorded, with concentrations, particularly in the Thames terraces (eg, Bridgland 1994; Wymer 1999) and the Solent (eg, Hosfield 1999; Ashton and Hosfield 2009; Harding et al. 2012). Important technological development occurred as the Levallois, prepared-core tools emerged in Britain during MIS 9.

This technology developed through MIS 8–6 (White et al. 2006). Evidence for sea level during MIS 9 is complicated by a lack of sites and appropriate dating material. However, in Purfleet sea levels were inferred at 14 m OD based on intertidal deposits. Although more recent work suggests that Purfleet was situated at the limit of tidal influence, some distance from the contemporary coastline (Schreve et al. 2002; Bridgland et al. 2012). Analysis of deposits assigned to the Purfleet interglacial indicates a range of habitats including riparian, woodland and grassland environments with climatic conditions that are thought to be warmer than the present-day (Bridgland et al. 2012). Sea levels were likely at, or above, present-day levels.

MIS 7 comprised three warm peaks interspersed by colder conditions and, as such, development of the British landscape during this period is complex. During the high stands, sea level was approximately 10 m below OD. During the cooler phases sea levels dropped, but by how much is open to debate (Pettitt and White 2012). Waelbroeck et al. (2002) suggest 25 m below OD and 85 m below OD during the two cold periods, respectively. From the close of MIS 7 there seems to have been a total population collapse in Britain; hominin groups appear to have abandoned Britain or suffered local extinction. It appears therefore, that Britain remained uninhabited until c. 40,000 years ago (MIS 3), although recently Wenban-Smith et al. (2010) have suggested possible occupation sometime between MIS 5d–5b. This abandonment is probably due in part to the breaching of the Weald-Artois Ridge (MIS 6) (Gupta et al. 2007; Toucanne et al. 2009). It may have been the case that access to Britain across the, almost permanently flooded, Channel was either impossible or only accessible during very brief periods after MIS 7 (see Ashton et al. 2011; Scott and Ashton 2011); this was partly driven by the increasing subsidence of the North Sea Basin. However, a broad range of flora and fauna, unlike hominins, did manage to colonise Britain, notably during MIS 5e.

Fluvio-glacial Sediment Deposition in the Palaeo-Yare

During the cold phases of MIS 10 and 8, when sea levels were lowered, glaciofluvial sands and gravels were deposited within the Palaeo-Yare system. Upstream, the Yare Valley Formation continued to develop and the Homersfield Terrace of the Waveney River was formed. Saalian deposits, classified as Unit 3 (a and b), were deposited and are identified within Area 240 and also throughout a large portion of the east coast aggregate block, where the remnants of this unit mark the limits of the floodplain.

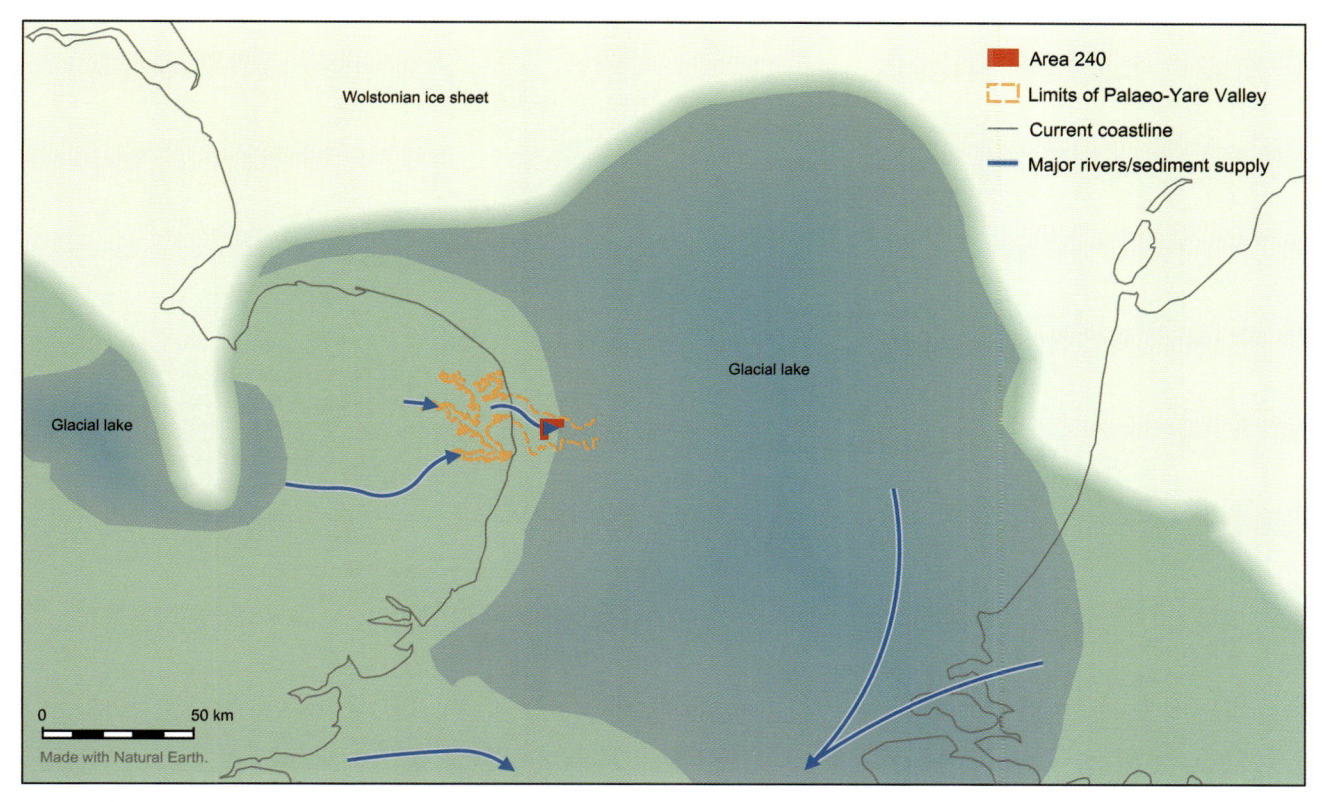

Figure 4.11 Generalised palaeogeography during the Saalian (MIS 6) (after Murton and Murton 2012)

Unit 3 (undifferentiated) is a complex sequence of sand and gravel deposits that form the infill of the late Anglian channel and broad floodplain deposit. In Area 240 the channel is orientated north-west to south-east across the Area and the southern edge of the channel is prominent; the channel edge cut is approximately 5 m deep. The northern edge of the channel is less obvious and is observed to be gently shoaling rather than the southern edge which is more of a steep cut.

Unit 3 is a complex sequence of deposits, not easily resolved in the geophysical data. The vibrocore data indicates lateral changes in sediment type within the unit and localised sequences which are not consistent, and as such not mappable, over the area. However, a distinct strong reflector is observed in the sub-bottom profiler data which marks a change in depositional environment or a hiatus of deposition within the unit. As such, Unit 3 can be divided into two broad units, Unit 3a and 3b.

Unit 3a is the deepest, and oldest, fill of the channel comprising coarse-grained sand and gravel and considered to have been deposited in a cold, glaciofluvial environment, although palaeo-environmental evidence from the vibrocores is lacking (Fig. 4.12). The age of Unit 3a cannot be stated with certainty; however given the post-Anglian development of the channel, sediments must date to younger than 434 ka and older than the overlying sediments dated to MIS 8. As the sediments were probably deposited in a cold environment a MIS 10 or late-MIS 12 date is inferred.

Unit 3b overlies Unit 3a in the channel feature and partially infills the channel. Beyond the limits of the channel the unit is extensive and forms a floodplain deposit throughout much of Area 240 and the aggregate block. In the west of Area 240 the underlying Unit 2 is intermittently observed sub-cropping the seabed sediments and no Unit 3b sediments are clearly observed on the geophysical data. This is likely due to recent dredging activity and the removal of Unit 3b sediments. The Unit generally comprises sands and gravels and is the target for aggregate dredging. As will be seen in subsequent chapters, it is this unit from which the hand axes were most likely to have been recovered.

Unit 3b is a complex unit of sand and gravel layers. The gravels are predominantly sub-angular to sub-rounded flint with notable amounts of quartz, quartzite, sandstone and occasional siltstone. Although gravel composition varies, there is no obvious discernible spatial pattern to the gravel content. Little palaeoenvironmental evidence was recovered from these sediments in VC2c (Fig. 4.6). The vibrocores within the ARZ (VC1c, VC2c and VC3c) indicate that Unit 3b comprises sand with occasional pebble layers and gravel near the top of the unit. To the north-east of this zone, situated to the south of the channel, vibrocore logs (eg, VC7c (Fig. 4.12)) indicate a more variable layering of fine-

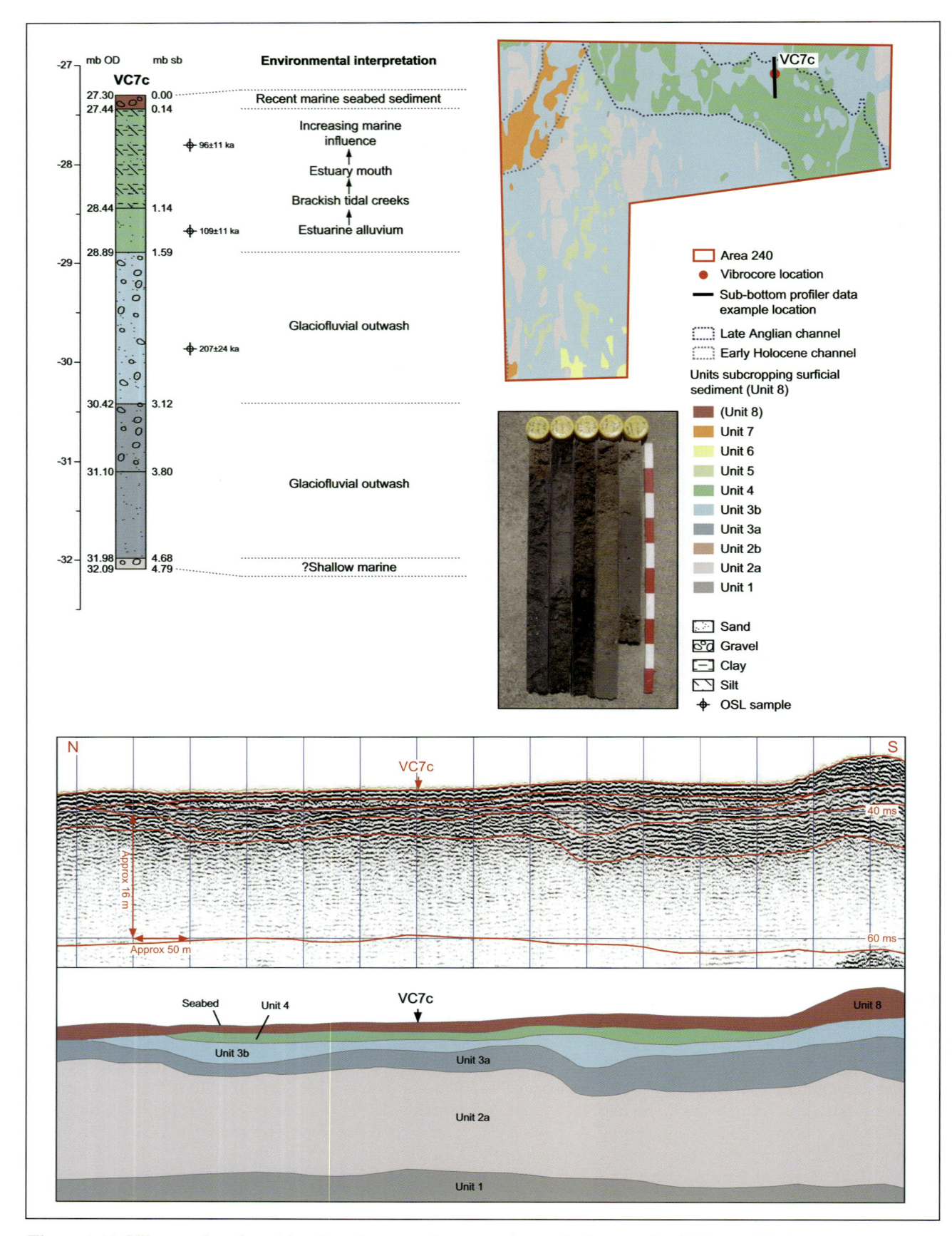

Figure 4.12 Vibrocore log, depositional environment interpretation and photograph of VC7c (with OSL dates from VC7b) with a seismic section illustrating the targeted channel infill sediments

Table 4.2 OSL ages of channel and floodplain deposits (Unit 3b)

Location/ vibrocore	Depositional unit	Depth (m below OD)	Age (ka); MIS	Reference
Area 240 (central floodplain)/ VC3b	Base of Unit 3b	31.0	418± 78; MIS 12/11	GL 10039; Wessex Archaeology (2011)
Area 240 (central floodplain)/ VC3b	Unit 3b below level from which handaxes were probably dredged	28.7	243 ± 33; MIS 7	GL 10038; Wessex Archaeology (2011)
Area 240 southern floodplain/ VC9b	Base of Unit 3b	31.6	283 ± 56; MIS 9/8	GL 10043; Wessex Archaeology (2011)
Area 240 channel/ VC7b	Upper channel deposits (Unit 3b)	29.9	207 ± 24; MIS 7	GL 10042; Wessex Archaeology (2011)
Area 254 (30 m north of Area 240)/ VC1	Upper Unit 3b sediment – bank structure	40.0	175 ± 23; MIS 7/6	Wessex Archaeology (2008c)
Area 319 (1.5 km west of Area 240)/ VC29	Unit 3b – base of bank structure	33.4	206.5 ± 29.5; MIS 7	Limpenny et al. (2011)
Area 319 (1.5 km west of Area 240)/ VC29	Unit 3b – middle bank structure	32.5	222 ± 28.7; MIS 7	Limpenny et al. (2011)
Area 319 (1.5 km west of Area 240)/ VC29	Unit 3b – upper bank structure	31.5	188 ± 19.7; MIS 7/6	Limpenny et al. (2011)

grained sands and gravel layers. This possibly indicates a change in flow regimes and deposition throughout the floodplain. It is likely that shallow braided channels were present within the floodplain as well as the main channel observed in the north of Area 240. There was no evidence of marine conditions or that the sediments were subject to marine reworking; evidence one would expect to recover if present. Given the geomorphology of the unit and lack of marine input these sediments are considered likely to be deposited in glaciofluvial conditions.

The glaciofluvial nature of Unit 3b indicates deposition during the cold conditions of the Saalian glaciation (MIS 8). This is largely confirmed by the OSL dating, however the precision of the dates resolves a window of MIS 9 to 7. Also, an OSL result of 418±78 ka (GL 10039) indicates possible deposition associated with late Anglian channel development prior to the Hoxnian inundation of the area. However, the age distribution of aliquots within this sample (ie, the dating protocol used for this project is based upon 12 sub-samples for each individual date) indicates mixing of significantly older unbleached or incompletely bleached material of equivalently pre-Anglian age (modes equivalent to MIS 15 and ~25/26) and with a significant modal distribution of younger possibly MIS 8 age. The subsequent averaging effect gives a date between MIS 12 and 11 (Wessex Archaeology 2013b; Toms 2011).

OSL dating of Unit 3b in the vibrocores indicates a likely deposition between approximately MIS 8 and MIS 6 (Table 4.2). Dredged sediments of Unit 3b within the ARZ were probably deposited prior to the warming limb of the glacial (MIS 8–7) (Table 4.2). Unit 3a sediments were deposited in a cold environment and are either contemporaneous with MIS 8 deposits or represent deposition in a previous cold stage (possibly MIS 10 or MIS 12).

Marine, fluvial and glacial depositional environments are demonstrably complex for the application of OSL sediment dating, particularly of this age (Rendell 1995; Harding et al. 2012). However, optical luminescence techniques can provide valuable chronological control for interpreting marine stratigraphy (Stokes et al. 2003). On balance, there does appear to be distinct modal distributions of aliquots suggesting the top of Unit 3b, ie, the active dredging surface was deposited by 250–200 ka during MIS 8–7 (Toms 2011) which confirms the interpretation of the overall site stratigraphy. Further optical dating assessments, perhaps incorporating single-grain techniques and/or high-resolution profiling measurements (Sanderson and Murphy 2010), would provide valuable information on the depositional history of these sediments and improve the clarity of the chronology.

Although there was little palaeoenvironmental evidence preserved within Unit 3b in Area 240, data from surrounding areas provide more information on the depositional environment. In the north-west area of Area 240 the southern edge of a bank feature is observed which extends into the northern aggregate dredging Area 254. This bank was studied as part of *Seabed Prehistory: Gauging the Effects of Aggregate Dredging* (Wessex Archaeology 2008c). The basal sediments of the bank generally comprise glaciofluvial sands and gravels and pollen suggest grasses and herbs dominate with lesser amounts of trees and shrubs, of which pine is most common.

To the west of Area 240 a bank three metres high, assumed to be of similar age and morphology to that in Area 254, was targeted by a vibrocore (VC29) as part of the East Coast REC research (Limpenny et al. 2011). It should be noted that the 'bank' feature is only described as such as because the floodplain was modified by later channel development in the early

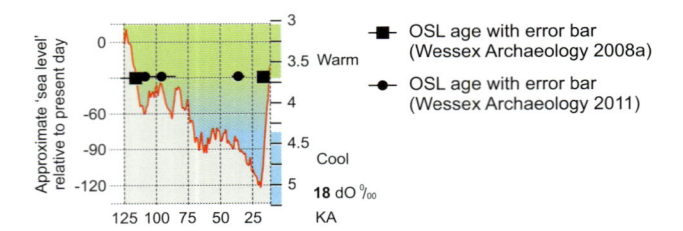

Figure 4.13 MIS curve for the Ipswichian (MIS 5e) and the Devensian (MIS 5d–2) illustrating OSL dating results for this period

Holocene, creating an erosional bank rather than a depositional bank feature.

The sediments comprise a basal shelly sand with a small assemblage of foraminifera and ostracods indicative of cold estuarine/shallow marine conditions. These basal sands were overlain by sand with interbedded silt and clay laminae with evidence of oxidation near the top of the sequence. A high abundance of molluscan remains was recovered dominated by shells of *Scrobicularia/Tellina* type and cockle (*Cerastoderma edule*) indicative of muddy, estuarine and intertidal conditions. This unit also contained a seed (*Ranunculus* type), charcoal, some fossilised wood and megaspores which are likely to have been reworked within these sediments. The vibrocore contained some singular reworked valves of shallow marine ostracods including species of *Cytheropteron*, *Semicytherura* and *Palmoconcha*. Well-preserved foraminiferal assemblages were recovered and were dominated by species of *Elphidium* and *Elphidiella* (*Elphidiella hannai*). These low diversity *Elphidiella* dominated assemblages are often indicative of pre-Anglian (MIS 12) cold, estuarine/shallow marine conditions (Funnell 1989). However, OSL dating indicates a younger age (Table 4.2).

As detailed in Table 4.2 the sediments from VC29 have been dated to between 222±28.7 ka at 32.50 m below OD and 188±9.7 ka at 31.50 m below OD (Limpenny *et al.* 2011). An uppermost date of 57±5.6 ka at 30.88 m below OD has at present been rejected as an age for deposition of the sediment on the grounds that it is possible that the sediment at this depth was exposed to sunlight during the Devensian (MIS 3) subsequent to its original deposition during the Saalian. The oxidisation noted within the sediments is a result of a fall in the level of the water table, suggesting that they have been elevated above sea level subsequent to their deposition.

Palynologically, the profile does not show full interglacial characteristics (MIS 7), but an early (Boreal) interglacial stage character, either associated with amelioration (MIS 8–7) or cooling (MIS 7–6). On-site vegetation habitat was initially one of grass-sedge fen which appears to have declined, with pine woodland dominant on higher ground.

To summarise, Unit 3b was deposited in a cold estuarine environment during MIS 8 but is probably associated with a cooling or warming limb rather than deposition at the height of the cold period.

During MIS 9, assuming sea level was close to present-day, one could expect the lower reaches of the Palaeo-Yare to be submerged or be intertidal and the upper reaches at least influenced by tidal regimes. During the high sea level it is likely that the till coastline to the north and south of the Palaeo-Yare continued to erode to some extent. During the warm phases of MIS 7 shallow marine conditions would prevail with more estuarine during the lowered sea level. No evidence of warm environment deposition associated with MIS 9 or 7 is clearly observed at Area 240.

The Last Interglacial: Ipswichian (MIS 5e; 130–115 ka)

The onset of the Ipswichian at MIS 5e was marked by an abrupt climatic transition from the end of MIS 6 with rapid melting of the glaciers and rapid sea-level rise (Fig. 4.13); global sea level rose to between 5 and 9 m above OD (Dutton and Lambeck 2012). On the Dutch coast sea level was, at most, a few metres higher than presently, although marine Ipswichian (Eemian) deposits lie at depths of -8 m (local datum) and greater, indicating the overall subsidence of the southern part of the North Sea Basin over the past 100 ka (van Gijssel and van der Valk 2005). The climate was similar to that of today, possibly slightly warmer with hot summers and mild winters (Barton 2005). Contemporary faunal remains discovered at sites such as Bobbitshole near Ipswich indicate a climate suitable for large mammals such as lion (*Panthera leo*), hippopotamus (*Hippopotamus amphibius*), straight-tusked elephant (*Palaeoloxodon antiquus*), rhinoceros (*Stephanorhinus hemitoechus*), giant deer (*Megaloceros giganteus*), red deer (*Cervus elaphus*), fallow deer (*Dama dama*), aurochs (*Bos primigenius*) and bison (*Bison priscus*) within floodplains (Wymer 1999). Evidence from Essex suggests the presence of vegetation comprising predominantly mixed oak forests (Allen and Sturdy 1980).

Surprisingly, this warm phase has not produced any certain evidence of occupation in Britain and recent re-evaluation of sites previously attributed to MIS 5e has been rejected (see Lewis *et al.* 2011). The conclusion is therefore that the artefacts were unlikely to be contemporaneous with the sediments in which they were found (Lewis *et al.* 2011; Pettitt and White 2012). The reasons for absence are debated but may involve several factors: for example, the island status of Britain during this time hindering access from

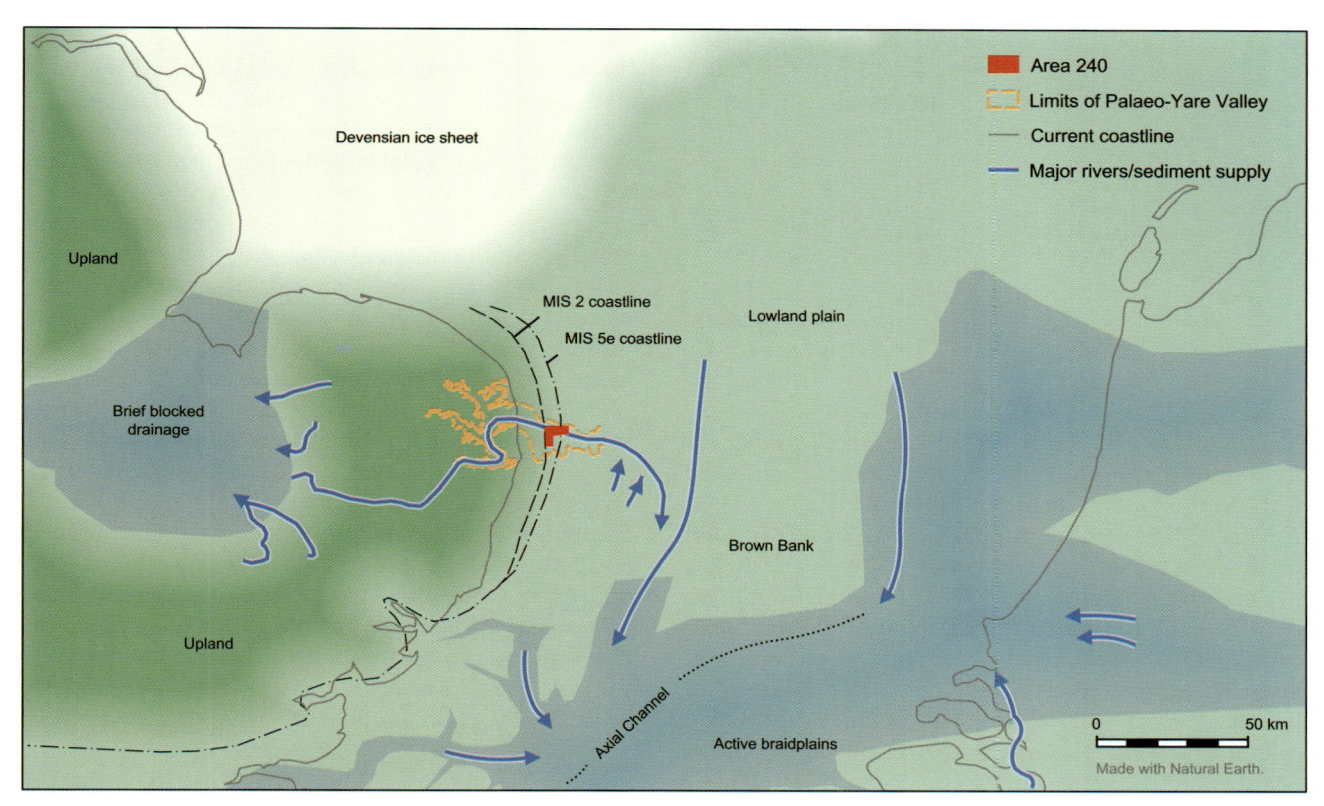

Figure 4.14 Generalised palaeogeography between the Ipswichian interglacial (MIS 5e) and the late Devensian (MIS 2) (after Limpenny et al. 2011; Hijma et al. 2012)

mainland Europe or that Neanderthals were better adapted to the mammoth-steppes of eastern Eurasia, rather than the dense forest of Western Europe (Ashton and Lewis 2012). However, the riverbank butchery site at Caours, Abbeville, France where the remains of rhinoceros, elephant and aurochs lie within deposits deposited c. 125,000 years ago (Antoine et al. 2003) and a MIS 5e site at Neumark-Nord in Germany (Roebroeks et al. 2011), indicate that human absence may be a local phenomenon (if not simply an absence of evidence).

Due to the significant sea-level rise associated with the Ipswichian, similar to the situation during Hoxnian interglacial (MIS 9), the influx of the sea on the Palaeo-Yare would have been limited by the till cliff which would have continued to erode during this time. There is little evidence of Ipswichian interglacial sediments remaining in the Great Yarmouth and Lowestoft areas with the exception of remnants of MIS 5e fluvial and semi-lacustrine deposits located in the upper reaches of the River Yare at Coston and riverine deposits in the Waveney Valley at Wortwell (Moorlock et al. 2000; Penkman et al. 2008; Lewis et al. 2011). Given the rise in the sea level during this time, much of the coastal areas would have been inundated. Of all the Pleistocene stages following the Anglian, only Devensian sediments are considered to be widespread at Great Yarmouth and its surrounding areas (Arthurton et al. 1994).

In the southern North Sea, situated north-east of Area 240, evidence of shallow marine Ipswichian deposits are observed in the form of Eem Formation (113 to 110 ka) comprising shelly sands, and to the west, becoming muddy sands and muds of a more intertidal nature (Cameron et al. 1992). However, similar to previous warm periods throughout the development of the Palaeo-Yare there are no obvious remnants of the marine incursion observed in deposits of Area 240.

Channel Re-activation and Continued Development: Devensian (MIS 5d–MIS 2; 115 ka–11.7 ka)

The Devensian was the last glacial stage to occur before the present climate amelioration. Between MIS 5d and MIS 2, ice sheets waxed and waned reaching their greatest areal extent by 27 ka (Gibbard and Clark 2011) with the southern extent of the ice sheet extending in a line from the Severn to The Wash. The sea-level curve for the Devensian reflects considerable climatic variability with long periods of relative cold and, overall, a general trend towards ever colder conditions, culminating in the last ice age. Area 240 would thus have been outside the limits of the ice but within the periglacial zone throughout the Devensian (Fig. 4.14). At the height of the Devensian, the water locked up in ice sheets caused a lowering of sea level to approximately 120 m below its current level and Britain would have been a peninsula of continental Europe throughout the majority of the period.

Early Devensian (MIS 5d–MIS 3; 115 ka–54 ka)

During MIS 5d–5a (110–75 ka) there was a general deterioration in climate and is characterised by interstadial (5c and 5a) and stadial (5d and 5b) periods. MIS 4 (*c.* 70 ka) marked the onset of very cold conditions in Europe with the Scandinavian ice sheet advancing into Denmark, Poland and the European Continental Shelf (Lowe and Walker 1997). At the MIS 5a–4 transition, sea level dropped considerably from approximately 25 to 90 m below present levels (Siddall *et al.* 2003). For at least parts of MIS 4 and the stadials of MIS 5 were similar harsh conditions to MIS 6 and conditions of polar desert and semi-barren tundra stretched across much of Britain (Pettitt and White 2012). However, pollen evidence in MIS 5d and 5b indicates that the harsh conditions did not limit tree growth altogether (Barton 2005).

The absence of evidence for hominin occupation since MIS 6 continues throughout the early Devensian. Recent work suggests that Britain may have been occupied as early as 100 ka (MIS 5d–5b), based upon two flint artefacts thought to be associated with a buried occupation level found at a site in Dartford, Kent and OSL dated to MIS 5d–5b (Wenban-Smith *et al.* 2010). However, further archaeological material and work on contextualising the finds would be required to prove occupation during this time at this site (Pettitt and White 2012).

During the early Devensian, as cooler conditions prevailed and the sea level lowered there was large-scale landscape reconfiguration on the southern North Sea plain. Major rivers formed and flowed to a lagoon situated to the north; there was only limited access to the open sea during late Ipswichian and early Devensian (Cameron *et al.* 1989). As the climate continued to cool these river systems would have dominated the landscape. Sediments infilling these river channels belong to the Brown Bank Formation and comprise up to 20 m of fluviatile current-bedded silt and finely laminated clays. The lagoonal sediments to the north comprise brackish-marine grey-brown silts which are extensively bioturbated and contain a thin layer of shelly gravelly sand towards the base (Cameron *et al.* 1989; 1992).

The upper reaches of the Palaeo-Yare would have continued to develop during the Devensian with at least some of the Yare Formation thought have been deposited during this period (Coxon 1979). As the sea level lowered at the start of the Devensian the channel in the north of Area 240 was reactivated and then infilled during the early Devensian (Unit 4).

Unit 4 is generally a fine-grained sequence up to 6 m thick. The unit is generally observed infilling broad shallow cuts. However, the unit also forms bank structures up to 3 m high. The unit is generally confined to the channel (Fig. 4.3); no remnants of early Devensian deposition are observed elsewhere within the floodplain in Area 240. The sediment sequence (based on two vibrocore profiles) indicates an increasingly saline profile. Based on the vibrocores (VC5c and 7c) two sub-units are identified: a lower unit of clayey, silty sand overlain by horizontally bedded sand and clay characteristic of tidal deposition indicating an increasingly saline profile. The lower of these units was palaeoenvironmentally assessed (VC7c; Fig. 4.12) and organic material was recovered from near the base of this unit and comprising fragments of stems and rootlets, possibly of brackish or estuarine plants, but equally could be marine algae (seaweed). Tasselweed (*Ruppia* sp.) seeds were also recovered; this plant is most commonly found in brackish cut-off pools on the coast or within lagoonal environments where there is both freshwater and marine input.

Foraminifera and ostracod species also indicated estuary environment consisting of brackish tidal creeks (*Cyprideis torosa*) giving way to a more estuary or estuary mouth environment (*Loxoconcha elliptica* and *Loxoconcha rhomboidea*). Molluscs indicative of estuarine and brackish water environments were also recorded including *Hydrobia*, both *ulvae* and *ventrosa*. There were also a number of shells of cockle (*Cerastoderma* spp.), of the Rissoidae family, of *Scrobicularia*/*Tellina*-type, saddle oyster (*Anomia ephippium*), oyster (*Ostrea edulis*) and a small scallop (*Chlamys*-type).

The pollen assemblage was dominated by pine (*Pinus sylvestris*), oak (*Quercus*) and grasses (Poaceae), with a consistent presence of elm (*Ulmus*), hazel (*Corylus avellana*-type), goosefoot family (Chenopodiaceae) and sedges (Cyperaceae). The preservation of pine was very good and suggests that it is unlikely to be derived from reworked material, a theory that is supported by the low presence of pre-Quaternary spores. A number of wetland pollen types were present. The presence of Chenopodiaceae and thrift (*Armeria maritima*) imply some marine or brackish influence. The abundance of deciduous woodland taxa also implies that this assemblage is likely to be from an interglacial/stadial warm period. Also present were bilberries and heath (*Vaccinium*-type) and heather (*Calluna vulgaris*) which may indicate areas of heath or mire in the local vicinity.

OSL dating within VC7b at 28.6 m below OD returned a date of 109±11 ka (GL 10037) and at

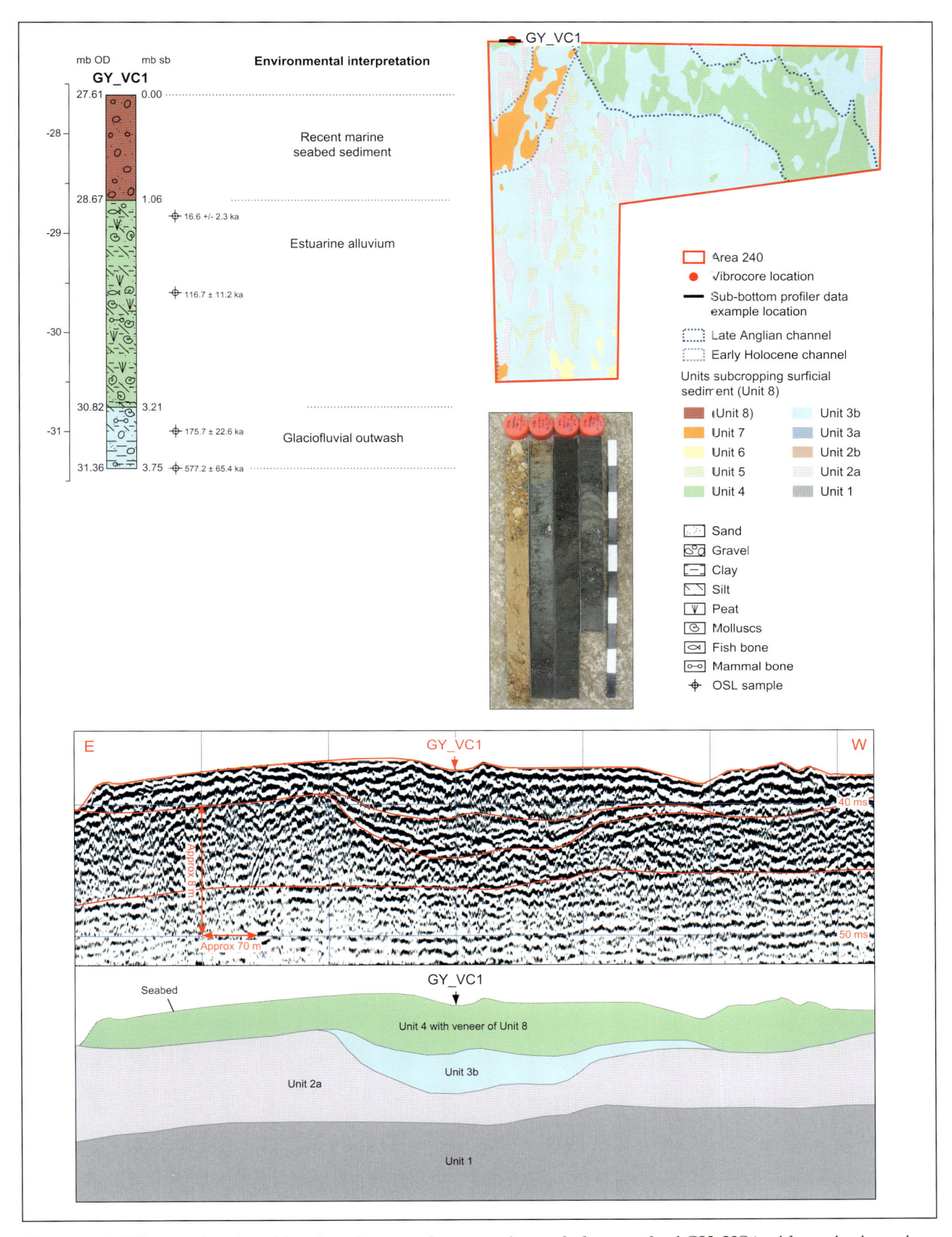

Figure 4.15 Vibrocore log, depositional environment interpretation and photograph of GY_VC1 with a seismic section illustrating Saalian floodplain deposits (Unit 3b) and overlying early Devensian bank deposits (Unit 4)

Figure 4.16 Reconstruction of the environment and landscape of Area 240 during the early Devensian. The estuary with shallow banks is surrounded by grasses and pine and birch woodland

27.8 m below OD returned a date of 96±11 ka (GL 10041) both correlating to the early Devensian (MIS 5c) and indicative of sediment deposition with a rise in sea level during the interglacial.

In Area 254 (situated directly north of Area 240), a similar sequence of organic freshwater sands and silts becoming brackish were OSL dated to 116.7±11.2 ka (Wessex Archaeology 2008a). These sediments are interpreted as Unit 4 with sediments infilling the deeper section of a shallow channel feature cutting into the underlying Saalian sand and gravel sediments (Unit 3b), as illustrated in Figure 4.15.

The environmental remains of this Unit 4 (GY_VC1) included a seed of the brittle water-nymph (*Najas minor*) and the pollen sequence observed is similar to that described above where pine and birch woodland is succeeded by oak and hazel. Some indications of increasing salinity were noted amongst the predominantly freshwater fauna indicative of a freshwater pool, lake or oxbow lake, and surrounded by a birch and pine woodland (Fig. 4.15). Figure 4.16 is a reconstruction of the environment and landscape of Area 240 during the early Devensian based on geophysical and geoarchaeological evidence.

Both sequences (in Area 240 and Area 254) are similar and, structurally, the infill sediments are interpreted to belong to the same depositional unit. The dates vary and suggest two possible scenarios for deposition. The first is that incision into Unit 3b occurred late in MIS 6 with infill sediments deposited during the early stages of the Ipswichian transgression. The second scenario involves incision during the early Devensian as the sea level fell at the start of MIS 5d and that the transgressive infill sequence is associated with the climate warming and sea-level rise during MIS 5c. Given the high sea levels in the area (up to 9 m above OD) during the Ipswichian (MIS 5e) compared to the slightly lowered highstand (approximately 30 m below present-day levels) during MIS 5c, the second scenario appears more likely.

There is no direct evidence of a relationship between the regional offshore Brown Bank Formation infilled channels as mapped by the BGS and the much smaller Unit 4 infilled channel within the Palaeo-Yare. The sediment compositions are similar as are the ages for deposition, and it is conceivable that they are contemporary. Evidence within the aggregate block, to the east, indicates that the presence of Brown Bank Formation is not as extensive as previously mapped. It is possible that the Palaeo-Yare channel acted as a tributary to these larger channels and evidence for this linking has been removed due to erosion, reworking or recent dredging activities. Alternatively it is possible that the Palaeo-Yare channel was cut off at this point. The large amount of accretion (up to 6 m) of Unit 4 would suggest stagnation of the channel or at least limited flow; this is supported by the fact that the unit seems to completely infill the channel.

Mid- to Late Devensian (MIS 3–MIS 2; 54 ka–11.7 ka)

MIS 3 (54 ka–25 ka) is typified by a sharply oscillating climate; short cooling episodes and milder climatic events are recorded. At the onset of MIS 3 sea levels rose to around 50 m below present levels (Waelbroeck *et al.* 2002) and then after 50 ka sea levels fluctuated between 60 and 80 m below present-day levels with a downward trend through time. In Britain, a peninsula of continental Europe, cool dry conditions encouraged the development of rich arid grasslands (mammoth steppe) which supported large mammals such as mammoth, woolly rhinoceros, lion, bear, etc. The dispersion of these animals probably also coincided with the recolonisation by Neanderthals. Evidence for hominin activity (Late Middle Palaeolithic: *c.* 60 to 35 ka) is sparse and suggests intermittent occupation for short periods between *c.* 60 and 41 ka with only a small number of finds associated with these sites (Pettitt and White 2012). The Late Middle Palaeolithic is generally dominated by hand axes supplemented by typical Mousterian tool forms in larger assemblages and, overall, corresponds to the continental Mousterian of Acheulean Tradition (MTA) which is present from MIS 5d onwards (Ruebens 2006). One of the few open-air (as opposed to cave) sites is at Lynford Quarry, Mundford, Norfolk where a Late Middle Palaeolithic lithic assemblage (595 artefacts, including 45 hand axes and debitage), and *in situ* mammoth bones, were found within organic fill deposits in an abandoned channel of the River Wissey (part of the Great Ouse valley). The infill deposits were dated to MIS 3 cutting into early Devensian sands and gravels (Boismier *et al.* 2003; Boismier *et al.* 2012).

In Europe, intermittent occupation is likely in the Netherlands from at least 50 ka. For example three bifacial leaf points of Mauern type and one unifacial Jerzmanovice point were recovered indicating either Neanderthals, anatomically modern humans or both (Rensink and Stapert 2005). In 2001, a portion of a *Homo neanderthalensis* skull, possibly belonging to a young adult male, was discovered in sediments extracted from the Zeeland Ridges, 15 km off the coast of the Netherlands (Hublin *et al.* 2009). The specimen was dredged up from sediments containing faunal remains (woolly mammoth, woolly rhinoceros, lion and hyena) and artefacts, including well-finished small hand axes and Levallois flakes. Although it was not possible to date the bone due to low collagen levels, the assemblage suggested a late Pleistocene, Late Middle Palaeolithic cold stage complex (Hublin *et al.* 2009), probably attributed to MIS 3.

The earliest direct evidence of *Homo sapiens* in Britain dates from *c.* 35 ka at Aurignacian sites such as Uphill in Somerset and Goat's Hole, Paviland (Pettitt and White 2012). This suggests a gap between occupation of *Homo neanderthalensis* and *Homo sapiens* of around 8,000 years. Environmentally, this could indicate Neanderthal abandonment for reasons other than competition and which may have taken place during a similar climatic and palaeogeographical scenario as the previous abandonment or population crash around MIS 7/6. However, taphonomic complexity with the material under study and the precision and accuracy of radiocarbon dating towards the limits of the techniques range (eg, Higham 2011, discussed in Pettitt and White 2012, 382) instil considerable uncertainty for establishing the timing and duration of Early Upper Palaeolithic activity in Britain (Pettitt and White 2012).

During the Early Upper Palaeolithic (*c.* 35 to 24 ka) occupation in Britain and in Europe was sparse and limited in time with the intense cold forcing populations to retreat to a few key areas before the Last Glacial Maximum (Housley *et al.* 1997).

The climate continued to deteriorate to a low point at 27–24 ka (MIS 2) – the Last Glacial Maximum – which corresponds with the maximum extent of the global ice volumes (Clark *et al.* 2009). The ice sheet extended as far as South Wales in the west and Lincolnshire in the east (Clark *et al.* 2012) *c.* 23 ka. Unglaciated areas were subjected to severe periglacial conditions with greater aridity, loess deposits and large areas subject to continuous permafrost. Lowe and Walker (1997) suggest that the southern North Sea was under extensive, continuous permafrost where lowstand sea level was between 114 and 35 m below present-day (Shennan *et al.* 2006). The wind-blown deposit of the Twente Formation is documented in the southern North Sea (*c.* 53°N) and is associated with these permafrost environments. These sediments were largely reworked during the Holocene transgression and are now only preserved as scattered outliers up to 1 m thick (Cameron *et al.* 1989; 1992).

The Devensian ice sheet indirectly modified drainage of the southern North Sea systems with the major rivers Thames, Rhine, Meuse and Scheldt flowing south into the channel via the Dover Strait. If the Palaeo-Yare floodplain was still active at this time, it is likely that it would have been a tributary to a south-flowing channel marked by the break in slope approximately 50 km from today's coast.

Throughout the Dimlington Stadial (31–16 ka) humans were absent until conditions had ameliorated significantly after 16 ka. Deglaciation occurred from 19 ka with more rapid and sustained amelioration around 16 ka. By this time large areas of Britain were free from ice (Clark *et al.* 2009; Clark *et al.* 2012). Fluctuations in temperature continued with warm periods (Greenland Interstadial 1/Windermere Interstadial) *c.* 14.7 ka as warm as today. Vegetation, however, would have been more continental with domination of woodland and grassland. The following cold periods (Greenland Stadial 1/Younger Dryas) saw a return to steppe and tundra environments (Hill *et al.* 2008) between *c.* 12.65 and 11.7 ka before the continued Holocene amelioration.

In the Paris Basin, hunter-gatherer activity is recorded from *c.* 16 ka (Gamble *et al.* 2004) and humans had re-established a presence in southern Britain by *c.* 14.7 ka with the Creswellian flint tool industry marking the start of the Late Upper Palaeolithic.

In the southern North Sea the active channels developed during the early Devensian would have continued to develop during the lowered sea levels of MIS 4 but with flow curtailed by fluctuations in sea level. OSL dating of the upper sediments of one of the Brown Bank Formation-filled channels (situated *c.* 27 km south-east of Area 240) indicated restricted shallow marine deposition of MIS 3 age (30,400±6,900 BP at 55.10 m below OD) (Limpenny *et al.* 2011). Within the lower reaches of the Palaeo-Yare there is no evidence of channel development after the early Devensian deposition of Unit 4. Throughout the majority of the channel, Unit 4 deposits have either infilled the channel or formed banks at the edges and within the remnants of the channel.

During the mid-Devensian it is considered unlikely that fluvial activity would have ceased completely in the downstream region of the Palaeo-Yare; the 'floodplain' terrace in the River Waveney is attributed a MIS 2 age and indicates continued activity upstream. The uppermost sediments of the Yare Valley Formation are attributed to the mid- to late Devensian (Cox *et al.* 1989; Arthuton *et al.* 1994).

It is possible the formation of the Unit 4 sediments blocked, or at least limited, flow through the channel and it is possible that smaller streams developed within the wider floodplain. The presence of small, isolated depressions in the wider floodplain may go some way to support this theory.

In the central and southern sectors of Area 240 small, infilled depressions are observed. The nature of the infilled sediments varies but, based on their structure in the geophysical data, they have been classified as a single unit (Unit 5). Predominantly the unit comprises slightly gravelly, slightly silty, fine- to medium-grained sand. The presence of clays associated with this fine-grained sediment indicates a low-energy depositional environment early in this sequence. Clay interbedded with shelly silty sand is observed in some cores and may be indicative of tidal rhythmic deposition suggesting an estuarine or near coastal depositional environment. With a continuously regressing coastline during the Devensian, an estuarine influence is more likely. The age of this unit is unknown and there is no stratigraphic association with any unit other than that it is younger than Unit 3b. It is likely that Unit 5 is Devensian in age; however whether it is early Devensian (Unit 4) or mid-Devensian (Unit 6) is difficult to state with certainty.

Unit 6 in the south of Area 240 is observed infilling broad shallow depressions in the surface of Unit 3b and is interpreted as alluvium. The infill comprises up to one metre of sandy gravel. Pollen evidence is sparse in this unit with some evidence of Poaceae and Cyperaceae, although they are represented in low numbers. An OSL date within Unit 6 in VC9b at 27.7 to 27.8 m below OD returned a date of 36±5 ka (GL 10045) indicating a mid-Devensian (MIS 3) age (Fig. 4.17).

It is possible that these sediments formed part of a wider, connected system of shallow channels that were subsequently eroded during the early Holocene transgression or have since been removed through the process of dredging in the area. Certainly, the nature of the Unit 5 sediments indicate their aggregate potential and the reduction in the lateral extents of the unit between the 2005 and 2009 datasets indicate removal of these sediments. These sediment units were not identified beyond Area 240 in the aggregate block; whether this was due to data quality or absence of features is difficult to determine.

Early Holocene Channel Development and Final Transgression (<11.7 ka)

Channel Development

The beginning of the Holocene in Britain saw a marked climatic amelioration accompanied by the transition from open tundra to birch and pine woodland within the general region (Allen and Sturdy 1980) and rising sea level (Shennan *et al.* 2012).

Broadly, the gradually shrinking lands within the region will have been occupied by seasonally mobile human groups who exploited a diverse range of resources and left behind many thousands of artefacts testifying to an organised and diverse lifestyle that spanned across the landscape. River valley sides and lakesides, bluff locations overlooking rivers with locally available flint sources, and locations close to spring lines, were particularly favoured (Limpenny *et al.* 2011).

Early Mesolithic sites and find spots are often found adjacent to wetlands and estuaries (Oxford Archaeology 2007), indicating a preference by Mesolithic communities for areas in which they could exploit the marine resources available in such environments. In the Early Mesolithic period, the

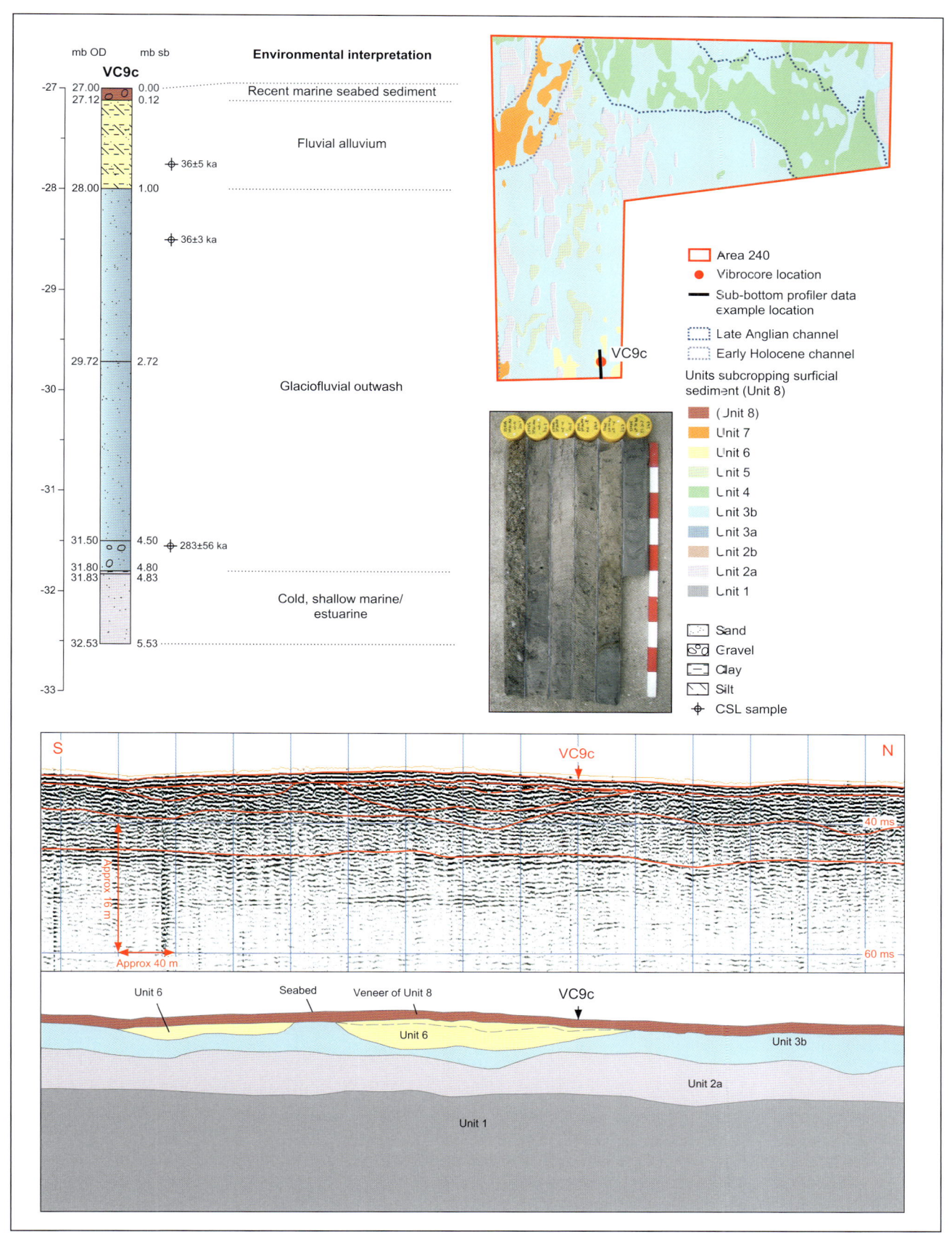

Figure 4.17 Vibrocore log, depositional environment interpretation and photograph of VC9c (with OSL dates from VC9b) with a seismic section illustrating the targeted floodplain deposits (Unit 3b) overlain by mid-Devensian sediments (Unit 6) infilling shallow depressions

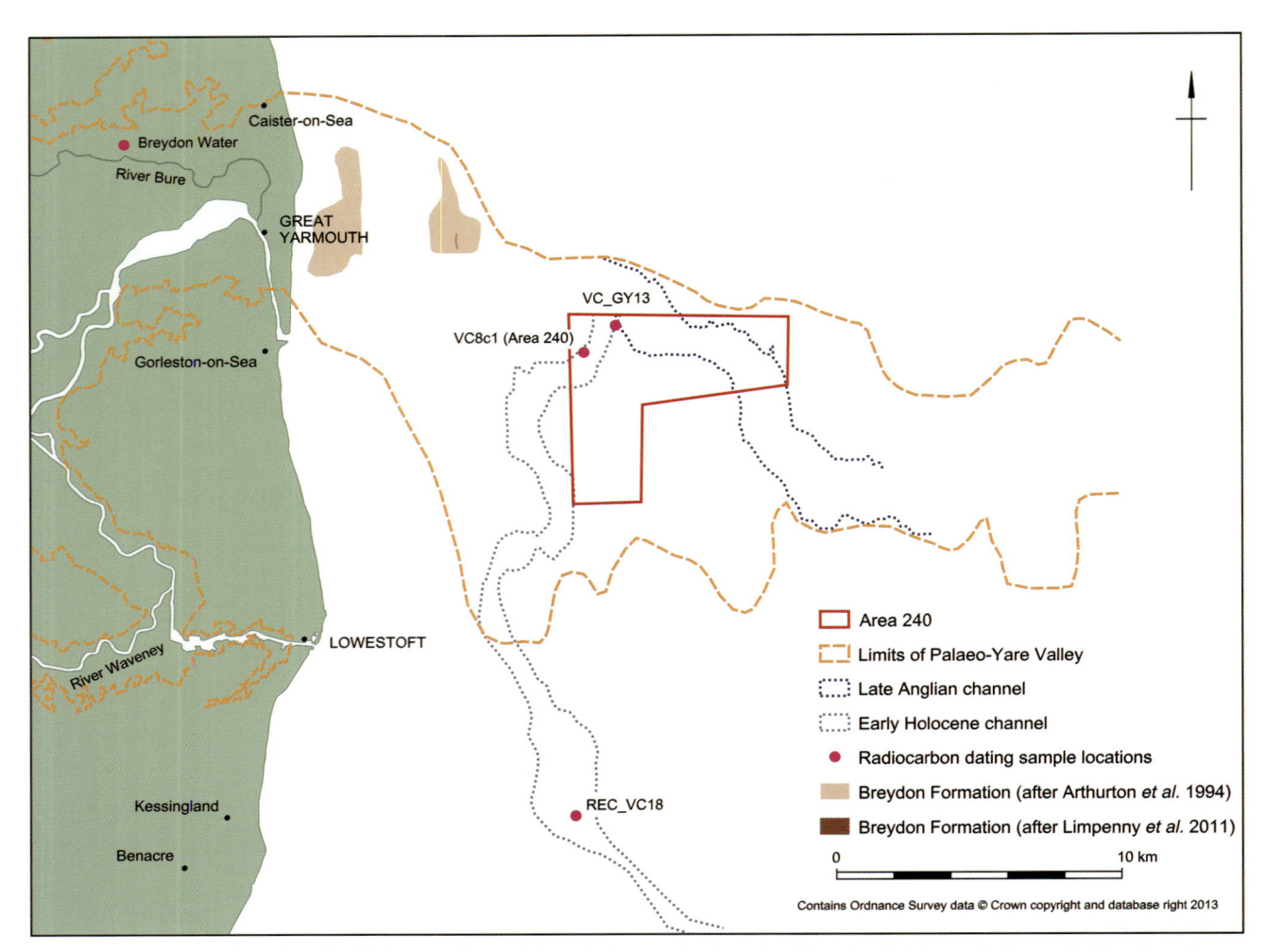

Figure 4.18 Location of early Holocene channel in relation to the late Anglian floodplain and channel. Core locations where sediments have been radiocarbon dated are illustrated

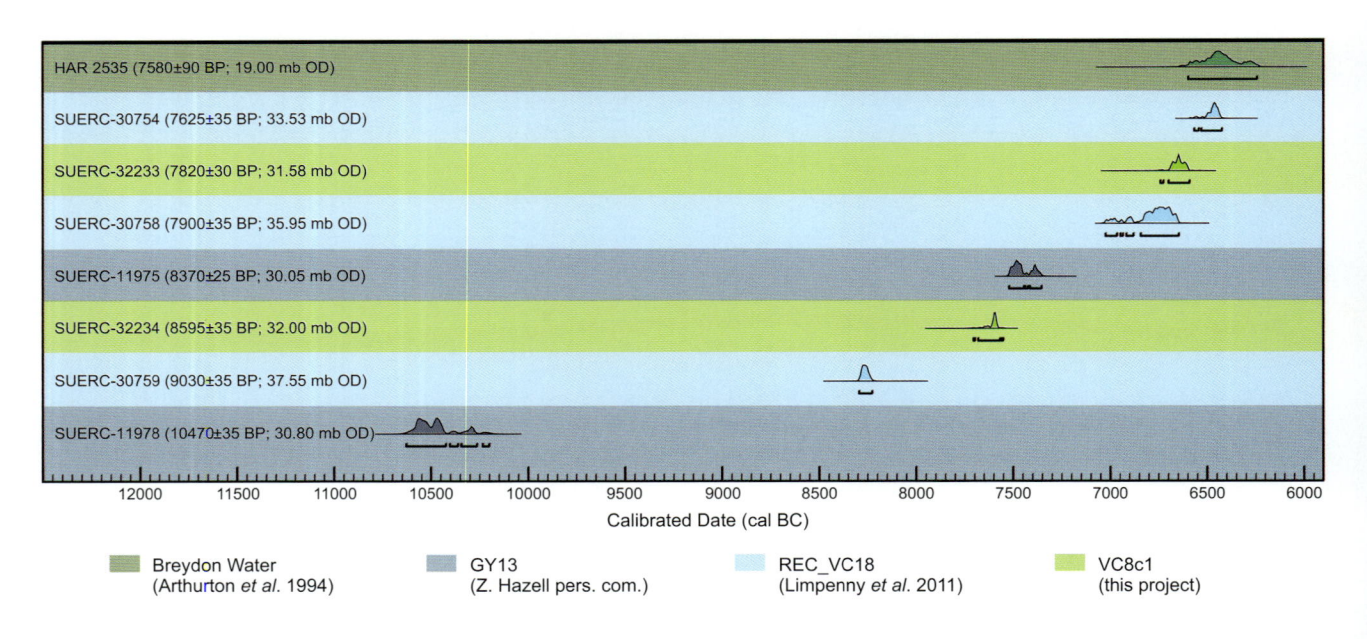

Figure 4.19 Radiocarbon dating results of a series of Unit 7 sediments

southern North Sea would have comprised undulating lowland drained by a complex of Pleistocene river systems. This extensive lowland would have been attractive for human occupation, not only providing access to both terrestrial and marine resources, but also enabling these Early Mesolithic communities to exploit the herds of red deer and other such mammals which migrated into Britain from the Continent as the climate ameliorated (Sumbler *et al.* 1996).

As the sea level continued to rise, brackish conditions would have begun to form within the river estuaries and the low-lying ground would have been subject to periodic flooding and the generation of marshland. Occupation sites within the area during the Late Mesolithic period are unlikely as the rising sea level would have progressively forced people further inland (*cf.* Leary 2011). However, Mesolithic communities continued to exploit the marine resources within the marshland until its full inundation. As such, there is considerable potential in these areas for the presence of archaeological material.

The now-submerged reaches of the Palaeo-Yare are likely to have been exposed as dry land until the start of inundation *c.* 8–7.5 ka (Shennan *et al.* 2000; Shennan and Horton 2002).

As the climate ameliorated during the early Holocene the development of the Palaeo-Yare was significantly modified. The confluence of the Rivers Bure and Yare at Great Yarmouth resulted in a large complex of alluvium, peat and fen silts adjacent to the coast. Peat of freshwater and brackish origins belonging to the Breydon Formation is a major component in the valleys of the River Yare and overlies the Pleistocene Yare Valley Formation gravels (Arthurton *et al.* 1994).

The Breydon Formation is thought to be preserved in the nearshore area off the coast of Great Yarmouth (Arthurton *et al.* 1994) although the lateral extent of the remnant sediments has reduced in size between the 1980s and 2009 (Limpenny *et al.* 2011), and indicated that the Palaeo-Yare continued to flow east from its known deposition. Approximately 8 km from the present-day coastline in Area 240 the remnants of a partially-filled channel within the Palaeo-Yare floodplain are observed (Unit 7) as they cut into underlying Unit 3b deposits. This early Holocene channel then continues to meander north–south for approximately 20 km before flowing south-east (Fig. 4.18). Based on the depth of the base of Unit 7 in Area 240 VC8c1 (32.7 m below OD) and the base of the unit in REC_VC18 at 36.9 m below OD, a gradient of 0.2 m km^{-1} is indicated. This is half as steep as the late Anglian channel of the Palaeo-Yare and such a deviation from the original floodplain is the first significant change in the downstream flow regime since its incision in the late Anglian. The diversion from the Palaeo-Yare may be due to the infilling and bank structures developed during the early Devensian (Unit 4).

Dating of the Unit 7 deposits and similar deposits in the downstream section of the channel has yielded similar ages to the lower deposits of the Breydon Formation. Radiocarbon dating of the peats indicate that Unit 7 is equivalent to the Basal Peat of the Breydon Formation (Fig. 4.19). Evidence of rolled and reworked peat in the offshore region indicates that erosion has continued since the last transgression. The Unit 7 sediment sequence indicates a progressively transgressive sequence from intertidal mudflat/saltmarsh deposited in the early/mid-Holocene. This is overlain by shallow marine/outer estuarine sand, which is in turn, is overlain by a shallow marine lag deposit and formed during the last transgression (Fig. 4.20). In the south of Area 240, the unit generally comprises silts and clays with occasional inclusions.

Unit 7 is only observed in the north-west and south-west of Area 240. However, it is likely that sediments are more extensive in the base of the channel but are not discernible in the geophysics data. Unit 7 is identified in the geophysics by bright reflectors due to the presence of peat and other organic matter in the sediment unit, not necessarily as a coherent mappable unit.

Unit 7 was sampled in VC8c1 (Fig. 4.20) and overlies the Unit 3b floodplain deposits. The unit comprises a fining-upwards sequence of estuarine gravely, silty and clayey sands from 32.70 to 32.06 m below OD, and contains ostrocods and foraminifera indicative of brackish environments dominated by the euryhaline ostracod *Cyprideis torosa*. A significant number of freshwater molluscs within the lower part of the unit were recovered including *Theodoxus fluviatilis*, *Bithynia* spp. and *Valvata* spp. These are all fresh-water species which thrive in moving water environments. The more abundant mollusc species indicated brackish and shallow marine environments, including *Hydrobia* spp. and *Cerastoderma* spp. Only one stray ostracod valve was recorded (*Candona* spp.) that can be attributed to non-marine environment and it is considered that these molluscs have been transported downstream from a freshwater habitat and incorporated into these estuarine sediments. It is interesting to note that the dominant ostracod *Cyprideis torosa* can tolerate widely fluctuating salinities from freshwater to hypersaline conditions, although its mass development is usually associated with a substrate consisting of soft mud and organic detritus in brackish salinities, between 2 and 16.5‰ (Meisch 2000).

These sediments are interpreted as deposition in an outer estuarine environment and is overlain by sediments deposited in an intertidal mudflat/

Table 4.3 Radiocarbon ages of Breydon Formation (Unit 7)

Location (vibrocore_year)	Depositional unit	Depth (m below OD)	Material	Radiocarbon Age (BP)	δ13C (‰)	Calibrated Date Range (95.4% confidence) cal BC	Laboratory code; reference
Breydon Water	Basal Peat of Breydon Formation	19	Not known	7580±90	-	6600–6240	HAR 2535; Arthurton et al. (1994)
Area 240 (VC8c1_2010)	Base of intertidal mudflat/saltmarsh deposit	32.06	*Phragmites* sp.	8595±35	-25.0	7710–7550	SUERC-32234; Wessex Archaeology (2011a)
Area 240 (VC8c1_2010)	Top of intertidal mudflat/saltmarsh deposit	31.57	*Phragmites* sp.	7820±30	-26.5	6750–6590	SUERC-32233; Wessex Archaeology (2011a)
Area 240 (GY13_2005)	Base of intertidal mudflat/saltmarsh deposit	30.80	Not known	10,470±35	-	10,630–10,200	SUERC-11978; Hazell (pers. com.)
Area 240 (GY13_2005)	Base of intertidal mudflat/saltmarsh deposit	30.05	Not known	8370±25	-	7530–7350	SUERC-11975; Hazell (pers. com.)
South of Area 251 (REC_VC18_2009)	Upper part of basal gravelly sands	36.65	Veneridae	9030±35	-4.0	8300–8220	SUERC-30759; Limpenny et al. (2011)
South of Area 251 (REC_VC18_2009)	Outer estuarine sediments	35.95	Organic sediment	7900±35	-23.3	7030–6640	SUERC-30758; Limpenny et al. (2011)
South of Area 251 (REC_VC18_2009)	Outer estuarine sediments	33.53	Organic sediment	7625±35	-23.0	6570–6420	SUERC-30754; Limpenny et al. (2011)

saltmarsh environment comprising horizontally bedded layers/laminae of sand, silt, peat and clay. The peat layers were up to 3 mm in thickness and appeared to comprise horizontally bedded plant remains (*Phragmites*). This is overlain by a sand unit up to 0.18 m with occasional small pebbles and marine molluscs, which is, in turn, overlain by a sandy gravel up to 0.32 m thick. At these levels the foraminifera (*Haynesina germanica*) and ostracod (*Hirschmannia viridis*, *Heterocythereis albumaculata* and *Loxococha rhomboidea*) faunas were indicative of a saltmarsh fringing an estuary which is likely to have developed within the tidal frame.

Additionally, within the intertidal deposit, several fragments of organics were recovered. Whole seeds of seablite (*Suaeda maritima*), a fragment of bud scale, a seed of white water-lily (*Nymphaea alba*), and one of probable pondweed (*Potamogeton* sp.) were also recovered. These samples also had fairly frequent small fragments of charcoal although none were large enough for identification. White water-lily is a freshwater aquatic, while pondweed is an aquatic that, although generally found within fresh-water environments, has some brackish-water species, living within estuarine habitats. Seablite is a coastal species common within saltmarshes and often found on mudflats. Ostracods were recovered that indicate littoral, sublittoral, marine and estuarine environments including *Hirschmannia viridis*, *Heterocythereis albomaculata* with *Cytherura gibba*. The foraminifera at this level were mixed, although also dominated by estuarine and marine forms, including species of *Elphidium* and *Miliolids*. Such an assemblage would be in keeping with rising sea level in the early Holocene and coastal/estuarine conditions during the formation of the deposit.

The pollen assemblage from Unit 7 was dominated by oak (*Quercus*) and hazel (*Corylus avellana*-type) with pine (*Pinus sylvestris*), elm (*Ulmus*) and grasses (Poaceae) present throughout. Other notable taxa present included birch (*Betula*), alder (*Alnus glutinosa*), small-leaved lime (*Tilia cordata*), Goosefoot family (Chenopodiaceae), meadowsweet (*Filipendula*), ribwort plantain (*Plantago lanceolata*) and Sedge family (Cyperaceae). The assemblage implies an area of mixed deciduous woodland with some stands of pine. There is a wetland environment present (indicated by taxa such as *Filipendula* and *Sparganium* sp.) which was subject to some marine/brackish influence, implied by the presence of Chenopodiaceae and sea-spurrey (*Spergularia*-type). The presence of sheep's sorrel (*Rumex acetosella*) and ribwort plantain, in association with peaks in the microscopic charcoal record, may imply some local disturbance, though this may be in the form of natural burning rather than being necessarily related to anthropogenic activity.

To the south-east of Area 240 (in Area 251) a peat sample was recovered and reported through the Marine Aggregate Industry *Protocol for Reporting Finds*

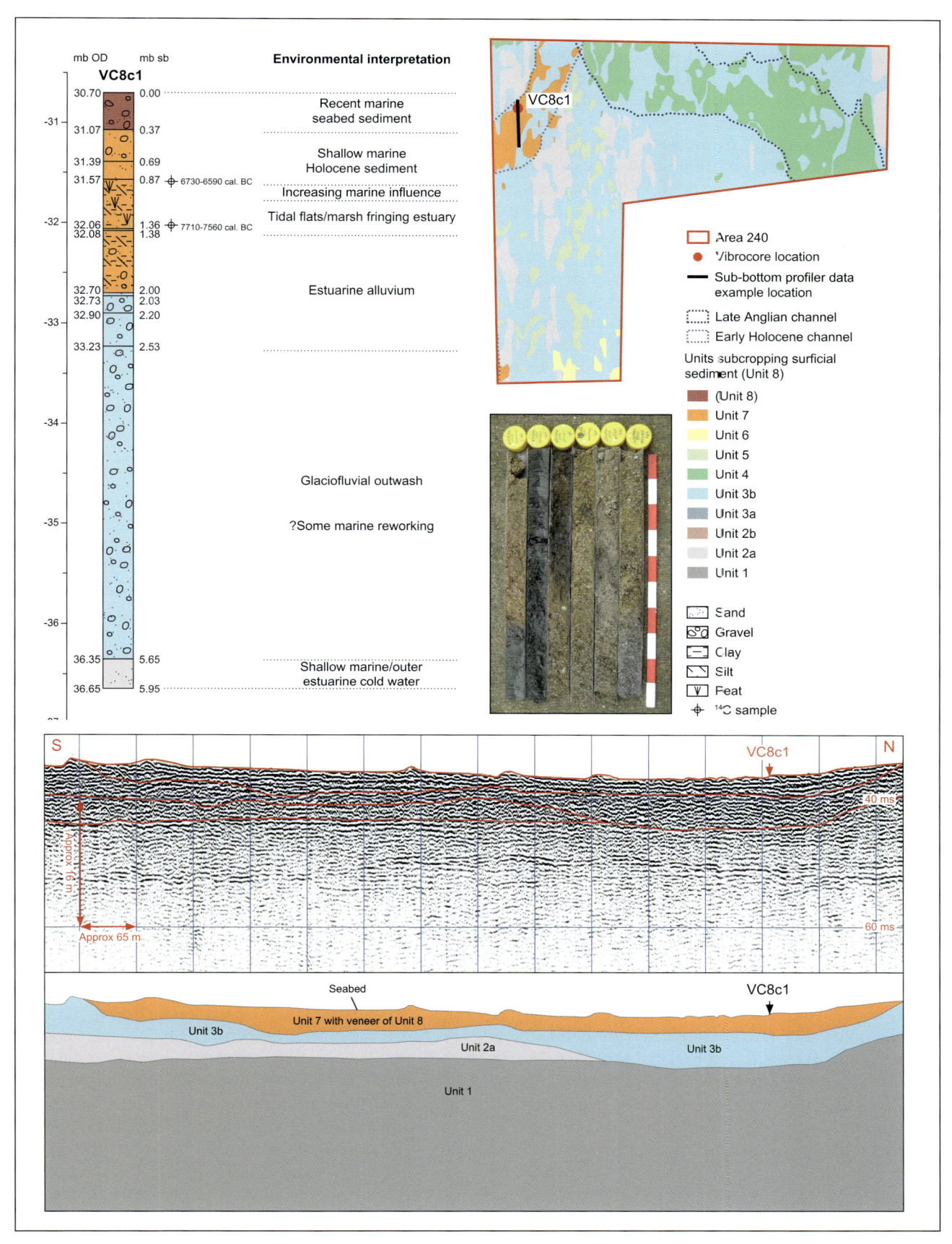

Figure 4.20 Vibrocore log, depositional environment interpretation and photograph of VC8c1 (with radiocarbon dates) with a seismic section illustrating the targeted Saalian floodplain deposits (Unit 3b) overlain by early Holocene channel deposits (Unit 7)

of Archaeological Interest (Cemex_0296). The location of the sample was fairly well constrained to a 1.4 km track within the channel. The peat sample was processed for the recovery and identification of waterlogged plant remains.

The peat sample from Area 251 contained relatively high numbers of seeds, as well as substantial numbers of wood fragments. The assemblage indicates the deposition in an ox-bow lake or similar cut-off chute. The presence of opercula of *Bithynia*, a mollusc species associated with flowing channels, suggests that some of the material may be derived from overbank flooding. The assemblage suggests the peat was formed on boggy ground, adjacent to a flowing river or stream, with only slight evidence for larger bodies of standing water. This evidence comes mainly in the form of seeds of white water-lily (*Nymphaea alba*) which is more common in pools, ponds and standing water in oxbows, but can be found in the still water of slow-flowing rivers.

The high presence of wood fragments may have been derived from shrubs adjacent to the peat, although the presence of seeds of sedge (*Carex* sp.), bogbean (*Menyanthes trifoliata*), common reed (*Phragmites australis*), marsh fern (*Thelypteris palustris*) and greater spearwort (*Ranunculus lingua*) points to some marshland and possibly wet grassland elements.

The presence of birch (*Betula pubescens*) seeds, aspen (*Populus tremula*) wood fragment and bud-scales, and willow (*Salix*) wood fragment (only small tree/shrub species), indicates probable open shrub woodland within close proximity to where the peat formed. This assemblage is typical of early Holocene assemblages. It might be noted that marsh fern is also recorded today in Britain in wet downy birch woodland (Rodwell 1991; plant community W4).

As inundation continued, Area 240 would have become a shallow coastal environment. Remnants of the till uplands to the north and south continued to be eroded by the encroaching seas and the alluvium and peats continued to develop in the upper reaches of the Yare Valley. The Breydon Formation Middle Peat formed between 4700±55 BP (3640–3360 cal BC, Q-2090) and 2170±55 BP (380–50 cal BC, Q-2086) in the Waveney valley (Arthurton *et al.* 1994). The base of the Upper Peat is dated to 1755±40 BP (cal AD 130–390, Q-2183) (Arthurton *et al.* 1994).

The palaeogeography and the environmental assessment certainly indicate an attractive environment for Mesolithic human activity. In the upper reaches of the Palaeo-Yare evidence of Mesolithic activity is represented by records of human remains from the marshy floodplain of the Yare between Surlingham and Strumpshaw. Lithic working sites appear to be concentrated close to Norwich and are associated with the margins of till and crag that overlook the margins of the river valleys. This trend may simply represent the eroding interfaces between sedimentary units which preferentially permit discovery of archaeological materials rather than a particular locational preference. Documented findspots of Mesolithic artefacts are more dispersed throughout much of the valley, with the notable exception of Kessingland. It has been noted that Norfolk, in general, has few excavated, or palaeoenvironmentally analysed, sites of Mesolithic date (Hill *et al.* 2008; Austin 2007).

Post-transgression Development

Although the post-transgressive processes are not of direct prehistoric palaeogeographic or archaeological interest they have nevertheless had an influence on the underlying Pleistocene sediments and are relevant in discussions of taphonomy and the preservation of archaeological material.

To the west of Area 240, nearshore, there are a number of large-scale sandbank features that measure up to 25 m high. D'Olier (2002) suggests that a large part of the source material was the erosion of the remnant till headlands that have continued to erode over the last 7000 years as the sea level increased, although erosion also occurred during previous transgressions. The total volume of sand within these banks is closely approximated by the volume of sand lost from the nearby Norfolk cliffs over the last 5000 years (Clayton 1989). Present-day sediment sources include the continuing erosion of the cliffline between Cromer and Happisburgh (the Anglian till of the Corton Formation) and are supplemented by a slowly eroding seafloor. The sandbanks off Great Yarmouth are surveyed regularly for shipping navigation purposes and illustrate the mobility of the sand within the thick deposits over the Palaeo-Yare floodplain (D'Olier 2002).

Post-transgression sediments (Unit 8) are observed throughout Area 240. The unit is observed as a veneer overlying the pre-transgression sediments or as bedforms, such as sand ripples or sandwaves that can measure up to 6 m high (Fig. 4.20). The bedforms generally comprise shelly well-sorted sand with occasional small (20 mm) oxidised mudstone inclusions, probably indicative of the units mobile nature. In between the sandwaves coarser sediments are observed comprising shelly, gravelly medium to coarse sand. Despite strong tidal currents the gravels have not been transported and are Pleistocene lag deposits with some inclusion of marine molluscs that indicate localised reworking. Further reworked elements were observed such as rolled peat and wood from underlying sediments, as well as recent inclusions of slag, glass and pottery, etc.

A number of broken iron oxide sheets were also recovered from grab samples and observed on photographs in the area. The sheets were pebble sized, up to 100 mm wide and a couple of millimetres thick and are thought to be reworked into the marine sediments. The orange colour is caused by iron oxide forming a crust around existing sediment. The sheets are friable and the centre of the crusts is composed of either hard fine-grained silts or very soft light grey clay. The mechanism for the formation of the sheets is unknown and could arguably be associated with shallow marine, intertidal or river environments. The fact that the material formed in a sheet-like deposit as opposed to clumps, may indicate some sort of laminar (smooth) flow and a sudden exposure to air that would allow fast drying, oxidation of the iron and the formation of sheets. This would be more likely to originate from a riverine/estuarine environment than a coastal intertidal area. This sort of deposition could occur within a floodplain of a river undergoing a flood whereby particles of fine material are transported to the floodplain, where water flow slows and allows the deposition of fine-grained silts and clays. As the flood waters recede the mud can form a crust during a period of baking. This would indicate pre-transgression formation, however, the depositional environment and age of these sheets remains speculative.

Within Area 240 the mobile sediments are reworked on a local scale with only limited input from the south. The sandwaves are mobile; however, comparisons between the 2005 and 2009 bathymetry data did not indicate major changes in the location or size of sandwaves.

There would have been reworking and redistribution of sediments during the last transgression, particularly during the period where the water was shallow enough that surficial sediments were affected by the wave base. Once the water depth increased reworking of sediments by geomorphic processes would likely have reduced. Modelling of present-day conditions indicate the potential depth of reworking for sand is less than 36.5 mm and gravel is up to 284 mm (Limpenny *et al.* 2011). As such, reworking predominantly affects the upper surficial sediments in the area but could indicate some reworking of the underlying units where gravel is the main component. Sediments buried greater than 0.3 m are no longer affected.

The sediment morphology and reworking of sediments is complicated somewhat by dredging over recent decades. Evidence of dredging is observed in the bathymetry data (Fig. 4.21) and is confined to particular areas, although dredging has occurred throughout Area 240 at varying times since 1993. The process of onboard aggregate screening, whereby unwanted sediment grades are returned to the water column and which with time settles onto the seabed in accordance with localised hydrographical conditions. Not only has material been removed from the seabed and sub-seabed but it has also been redeposited. This not only has implications for sediment deposition models but also has an impact when assessing potential context for archaeological material on and sub-seabed.

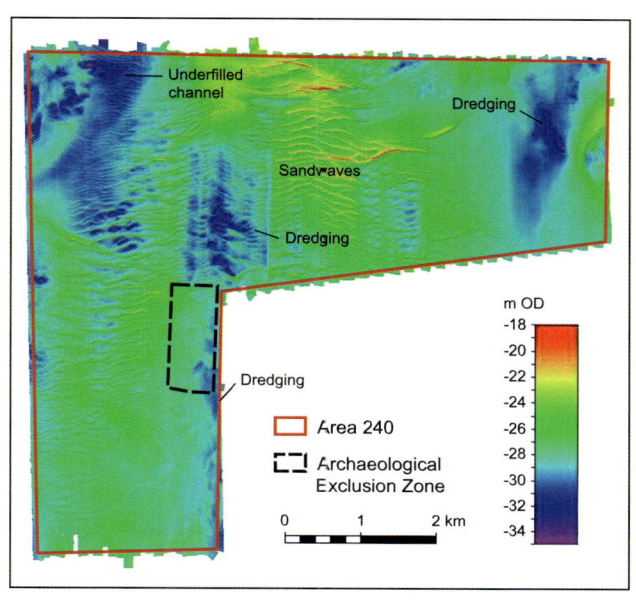

Figure 4.21 Multibeam bathymetry illustrating natural bedforms including underfilled channels and sandwaves up to 6m high and the effects of dredging

Preservation of Sediments in Area 240

The Palaeo-Yare Valley initially formed at the end of the Anglian and then continued to develop through to present-day. The Yare Valley has been active since its inception to the present-day throughout a number of oscillations from glacial to interglacial conditions and associated transgressions and regressions. During cold periods subsequent to the formation in the late Anglian, when sea levels were lower, sands and gravels were deposited on the valley floor and formed terraces in the upper reaches. During warm periods the sea levels rose and the Yare Valley became increasingly affected by tidal conditions; in its lower reaches it became an estuary and in the eastern extremes, shallow marine conditions prevailed.

The investigations of the sediments of Area 240, and aggregate block, indicate a complex sequence of sediments dominated by cold environment deposition that have been deposited and redeveloped over numerous glacial cycles, although evidence of some warmer environments did persist. At each phase of development sediments associated with the lower reaches of the Palaeo-Yare would have undergone reworking. However, although evidence of reworking is apparent, as indicated by sediment mixing or the presence of reworked pollen and ostracods assemblages, reworking was not wholesale and there has been a high level of preservation of the sediments.

Ice sheets arguably cause the greatest amount of destruction and reshaping of the landscape, as seen by the diversion of the Thames and the partial destruction and burying of the Bytham River by the Anglian ice sheet. Since its initial development at the end of this glaciation the Palaeo-Yare has been free of ice in subsequent glaciations. Although subjected to periglacial conditions and the associated small-scale reworking of sediments, the river system has been free from major remodelling.

Since their original deposition, the channel and floodplain deposits have undergone four possible episodes of transgression and effects by coastal and shallow marine processes: Hoxnian (MIS11), mid-Saalian (MIS 9) when sea levels were similar to present-day and Ipswichian (MIS 5e) when sea levels were approximately 5 m above present-day and, finally, the Holocene transgression. Even accounting for isostatic shifts, Area 240 would have been inundated during these periods and previously deposited sediment units would have been subject to tidal currents, wave action and associated reworking. The degree to which these processes would have influenced the environments would depend on the pace of the transgression as dictated by the glacioeustatic conditions as well as local factors such as wave formation (height and fetch), coastline topography and shoreline configuration.

In the case of Area 240, and in fact all submerged areas, it is difficult to know how much sediment has been eroded (or the amount of reworking and redistribution of sediment over a regional scale) during any one transgression; therefore there remains a question as to the degree of impact on the pre-transgression sediments. The region of the submerged Palaeo-Yare appears to have always been a low gradient estuary/coastal shelf and therefore the impacts of waves may have been minimised. In MIS 11, 9 and 5e it is considered that the coastline configuration with eroding till high grounds to the north and south may have protected some sediment erosion in the Palaeo-Yare floodplain, but also may have increased deposition of the eroded till during these periods. During these periods it may be that coarse-grained sediment was less affected than unconsolidated finer-grained sediments unless they were already buried. Also, the rate of coastal erosion would have guaranteed ample replenishment and high seabed sediment loads providing additional preservation of the underlying deposits.

The last marine transgression had an impact that can be seen in widespread reworking of sediments which now form the marine uppermost unit. This is formed from underlying sediments and major erosion of an estimated 6 km of till coastline. However, in Area 240 and the surrounding area there is evidence that the last transgression did not completely rework earlier landscapes, for example, the preservation of the partially infilled early Holocene meandering channel. Also, to the south of the area off Lowestoft, remnant intertidal deposits are observed (Moorlock *et al.* 2000; Limpenny *et al.* 2011).

Although it is clear that floodplain and channel deposits have survived these numerous transgressions, it is noted that there are no typical terrace deposits observed in Area 240 and the form of the channel does not conform to the typical terrace development (*cf.* Bridgland 2002). This could be due to a number or combination of factors. Firstly, terraces are unlikely to survive in areas of no uplift. Although the upper reaches of the Yare Valley exhibits some uplift indicated by the Homersfield (MIS 8) and Broome (MIS 6) terraces, these converge towards the present-day coastline, the location of a 'hingeline' with uplift to the west and subsidence to the east. This effect of the hingeline is also apparent in the development of the major pre-Anglian river terrace development (Rose 2009; Westaway 2009). The degree of subsidence in Area 240 is unknown but given the proximity to the 'hingeline' it is considered likely to be minimal.

In a channel not experiencing uplift, such as the lower reaches of the Palaeo-Yare, these sediments are reworked over time rather than being preserved in stacked sediment sequences (Rose 2009). In Area 240 it is likely there is considerable reworking of sediments from the initial incising of the river and deposition of coarse glaciofluvial material at the late Anglian (MIS 12) to the Saalian glaciation (MIS 6) and no warm period sediment deposition is preserved, due to erosion and removal or reworking. Estuarine alluvium deposited in the early Devensian (MIS 5c) was not subsequently reworked or significantly eroded and indicated a change in the flow regime during the Devensian.

Perhaps as important in terms of the preservation of Middle Pleistocene and early Holocene sediments is the impact of dredging activity over the past decades. In the aggregate block the volume of the Palaeo-Yare floodplain and channel deposits removed is unknown.

The target for the aggregate is the cold environment sands and gravels (Units 3b, 5 and 6) rather than the more temperate finer-grained sediments (Units 4 and 7). Since 1993 Electronic Monitoring System (EMS) data record information on date, location and time spent dredging. However, it is difficult to calculate the volume of sediment that this equates to as depth of dredging remains unknown. As such, it is difficult to estimate the size of the Palaeo-Yare prior to dredging. The 2005 bathymetry dataset shows obvious evidence of heavy dredging and comparison between the 2005 and 2009 data indicate up to one metre difference in areas between sandwaves where the seabed is not mobile under hydrodynamic conditions. The depth changes are probably due to dredging.

In the ARZ there is evidence of dredging activity, and obviously without dredging the flint artefacts would never have been recovered. However, subsequent sampling of the seabed has revealed that despite a number of years of dredging activity in Area 240, archaeological material has not been completely removed and the sediments in which the archaeological material is buried remain preserved. The presence of submerged sites and associated *in situ* archaeological material within these environments is dependent on the nature of the sediment enclosing the archaeological material and the sub-seabed depth of the artefacts. These issues, with regards to the preservation of archaeological material recovered from Area 240, are discussed in detail in Chapter 6.

Chapter 5
The Continued Search for Archaeological Material

Introduction

Large numbers of stone, typically flint, early prehistoric artefacts and faunal remains have long been found in sediments associated with river channels (Wymer 1999; Pettitt and White 2012), either in modern floodplain sediments or relict floodplains preserved above the modern valley floor following periods of uplift and river incision (Bridgland 1994; Bridgland and Westaway 2008; Lewin and Gibbard 2010). As discussed in the previous chapter, remnants of these river systems exist in now-submerged areas and have potential for archaeological material. Finds recovered from river terraces may be made and left *in situ* or material may be eroded and reworked from primary contexts and incorporated into other sedimentary deposits such as river floodplains further downstream (and subsequently terraces if further incision and/or uplift occur) (Wymer 1999). In a particular location – onshore or (now) offshore – there may then be a mixture of archaeological materials of various periods, and of various provenances, within preserved sedimentary units.

The recovery of Palaeolithic artefacts and Pleistocene faunal remains from the southern North Sea has a long history predominantly associated with the fishing industry and, more recently, the dredging industry (Godwin and Godwin 1933; Glimmerveen *et al.* 2004; Mol *et al.* 2006; Hublin *et al.* 2009). Numerous mammal remains have been reported from a relatively restricted area in the southern North Sea between the Brown Bank area and the Norfolk coast, which have yielded Early and Middle Pleistocene mammal fossils (van Kolfschoten and Laban 1995; De Wilde 2006; Strijdonk *et al.* 2011; 2012). Isolated finds of artefacts such as flints, bone spearheads, and reworked or carved fossil mammal bones are also documented (Long *et al.* 1986; Coles 1998; Flemming 2002). Finds and faunal remains have been, and continue to be, reported from North Sea aggregate dredging areas via the Marine Aggregate Industry *Protocol for Reporting Finds of Archaeological Interest* (BMAPA and English Heritage 2005) (Fig. 5.1). Also, the presence of intact archaeological sites on the French channel coast at Fermanville (Scuvée and Verague 1988), discovered by petroleum geologists in 17 m water, also highlight the often fortuitous nature of discovering submerged sites.

Predominantly these finds are isolated and have little or no contextual information such as spatial location or vertical stratigraphic context. The reporting and knowledge of these finds is important for interpreting the general archaeological potential of an area; however, such finds do not necessarily indicate the presence of preserved archaeological sites. There are relatively few examples of known submerged Palaeolithic and Mesolithic sites worldwide. Predominantly, the known, well-investigated sites are situated in relatively low-energy environments in nearshore coastal areas and are accessible for diving investigations. For example, in Danish waters diving investigations have proved successful in locating Palaeolithic and Mesolithic sites in water depths less than 20 m (eg, Andersen 2011; Uldum, 2011; Fischer 2011). These sites were investigated primarily by divers, although heavier equipment such as industrial sand-pump dredgers and hydraulic digging machines were also employed (Fischer 2004). Archaeological investigation by divers at Bouldnor Cliff in the western Solent, England, led to the discovery of Mesolithic occupation sites in water depths less than 20 m (Momber *et al.* 2011). Assessing sites and sampling for artefacts at offshore (versus coastal) sites can be demonstrably complex and challenging.

Sampling for archaeological material can be problematic whether in a terrestrial or marine environment. Environmental factors specific to a site under investigation need to be taken into account and methodologies designed accordingly. Referring to terrestrial work, survey methods are dependent upon abundance and clustering, obtrusiveness, visibility and accessibility of a study area (Schiffer *et al.* 1978). It was acknowledged in previous studies that '*archaeological research of any sort that uses survey data from regions where site discovery is difficult must confront and resolve, or at least acknowledge, discovery problems*' (McManamon 1984; 223). This also applies for surveys in the marine environment, where principal environmental conditions under consideration include, but are not limited to, water depth and turbidity, hydrodynamic conditions and sediment stratigraphy.

Unlike nearshore archaeological sites, such as at Bouldnor Cliff, Area 240 is not conducive to diving methodologies; water depths approaching 30 m mean that diving time would be limited using standard air

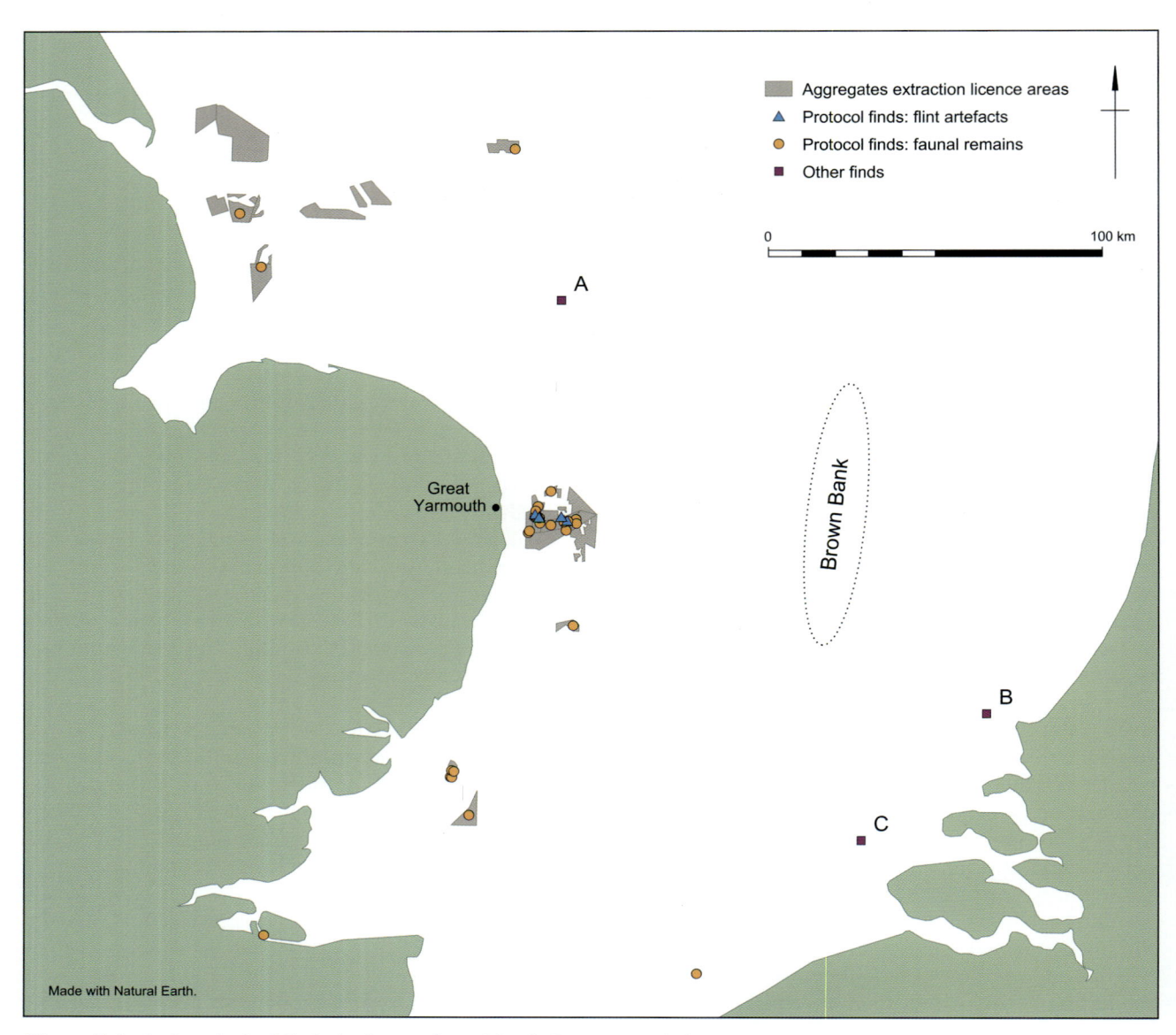

Figure 5.1 Archaeological finds in the southern North Sea reported through the protocol and (A) Late Glacial antler point from Leman and Ower Bank (Godwin and Godwin 1933), (B) large number of faunal remains recovered from Eurogeul shipping lane (Mol et al. 2006), (C) Neanderthal skull fragment (Hublin et al. 2009)

(or would require mixed-gas/technical diving to work effectively at such depths). Even more importantly, the strong currents in the area would mean that diving could only be conducted at slack water (the state of the tide when it is turning), which further limits diving operations. Finally, the visibility in the area is notoriously poor, further hindering a diver's ability to work. As a result, the prospect of locating flints, particularly in an area of 3.5 km², without a more precise location for the findspot, is remote and would require a major commitment of both time and financial resources. Cost effectiveness would be low and the chance of failure to find artefacts would be high. Although considered as a potential methodology, ultimately diving was discounted for this project. The size of the area and the environmental conditions favour the use of seabed sampling techniques.

Remote Sampling Techniques

Sampling techniques are systematically and routinely used in the aggregate industry as part of benthic studies in preparation for the marine ecological assessment element of an Environmental Impact Assessment (EIA) and as part of aggregate resource assessments. Typical ecological sampling techniques comprise trawls, dredges (anchor and towed), grabs and corers. Optical systems are also routinely used as a valuable, non-destructive method for assessing seabed environments (Cefas 2008).

Plate 5.1 The Hamon grab deployed from the vessel to sample typically 10–15 litres of seabed sediment

The use and development of these existing methodologies for archaeological applications is benefitted by techniques whose routine deployment is already well established and have the potential to enable integration of archaeological surveys with other sampling studies over the course of aggregate dredging and renewable licensing. Of the techniques applicable to recovering archaeological material, such as flint – only grabs, have been previously trialled for archaeological purposes, the results of which are described below.

Grab Sampling for Artefacts

Grabs provide discrete samples of the seabed. There are a variety of grabs available and the selection of which is generally dependant on the targeted seabed sediment, and the quantity required. Soft sediments (such as mud and silt) can be targeted by a wider range of equipment than coarse sediments (sands and gravels) (Cefas 2008). Typically, the two types of grab sampling equipment used in areas of coarse sediment are the Hamon grab and the hydraulic clamshell grab.

The Hamon grab acquires samples of 10–15 litres (Pl. 5.1) and is widely used in both habitat mapping and biological monitoring surveys. This is the recommended tool for sampling the benthic macrofauna from coarse seabed sediments (Boyd 2002). The Hamon grab was extensively trialled as an archaeological tool in the Seabed Prehistory surveys in the Arun, East English Channel and Humber surveys (Table 5.1).

A gridded sampling strategy was adopted for these surveys in order to achieve a systematic distribution of samples. Samples were acquired based on a 100 m² grid arranged overlying palaeogeographic features of interest identified in geophysical data. The systematic strategy was adopted to allow an appraisal of the efficiency of the sampling methods as an investigative archaeological technique. The gridded strategy would also assess any discernible correlation between the presence, or absence, of artefacts and the geoarchaeological features which had been identified previously by geophysical and vibrocoring techniques.

At each location a single grab sample of 8 to 10 litres was acquired and was wet sieved to >1 mm fraction through a sieve mesh on board. Subsequent sieving in the laboratory was undertaken through 10 mm, 4 mm and 1 mm mesh sieves. Each size fraction was then dried and sorted to identify artefacts of archaeological relevance. Probable struck flint, faunal remains and environmental remains (peat, plant, wood, etc.) were sampled with the Hamon grab, plus modern material such as slag, clinker, ceramic building material and glass.

Flint artefacts were recovered using a Hamon grab and a systematic sampling strategy from the Palaeo-Arun survey area associated with a wide palaeovalley (Fig. 5.2). The flints showing the greatest anthropogenic potential were assessed on the presence of bulb, striking platform, and the overall nature of the flint. Flints classified as probable artefacts contained elements which appeared to have a possible function, or which had elements that are not easily ascribed to natural processes alone, such as

Table 5.1 Summary of previous surveys using Hamon grabs for sampling archaeology

Study area	Date	Size of grab sample survey area (km)	Number of grab sample targets	Volume sediment (metric ton)[1]	Recovered flint	Reference
Palaeo-Arun	September 2003	1x1	100	0.202	3 highly probable; 4 probable; 18 possible	Wessex Archaeology (2008a)
Palaeo-Arun Additional Grabbing	June 2005	2x4 (including 1km² sampled previously)	400 (500 including previous sample targets)	0.808 (1.010)	12 highly probable; 10 probable; 19 possible	Wessex Archaeology (2008b)
Eastern English Channel	September 2005	0.5x2	100	0.202	No worked flint recovered	Wessex Archaeology (2008d)
Humber	July 2006	1x1	100	0.202	No worked flint recovered	Wessex Archaeology (2008e)

[1] based on average mass of 2,020 kg/m³ for sand and gravel sediments

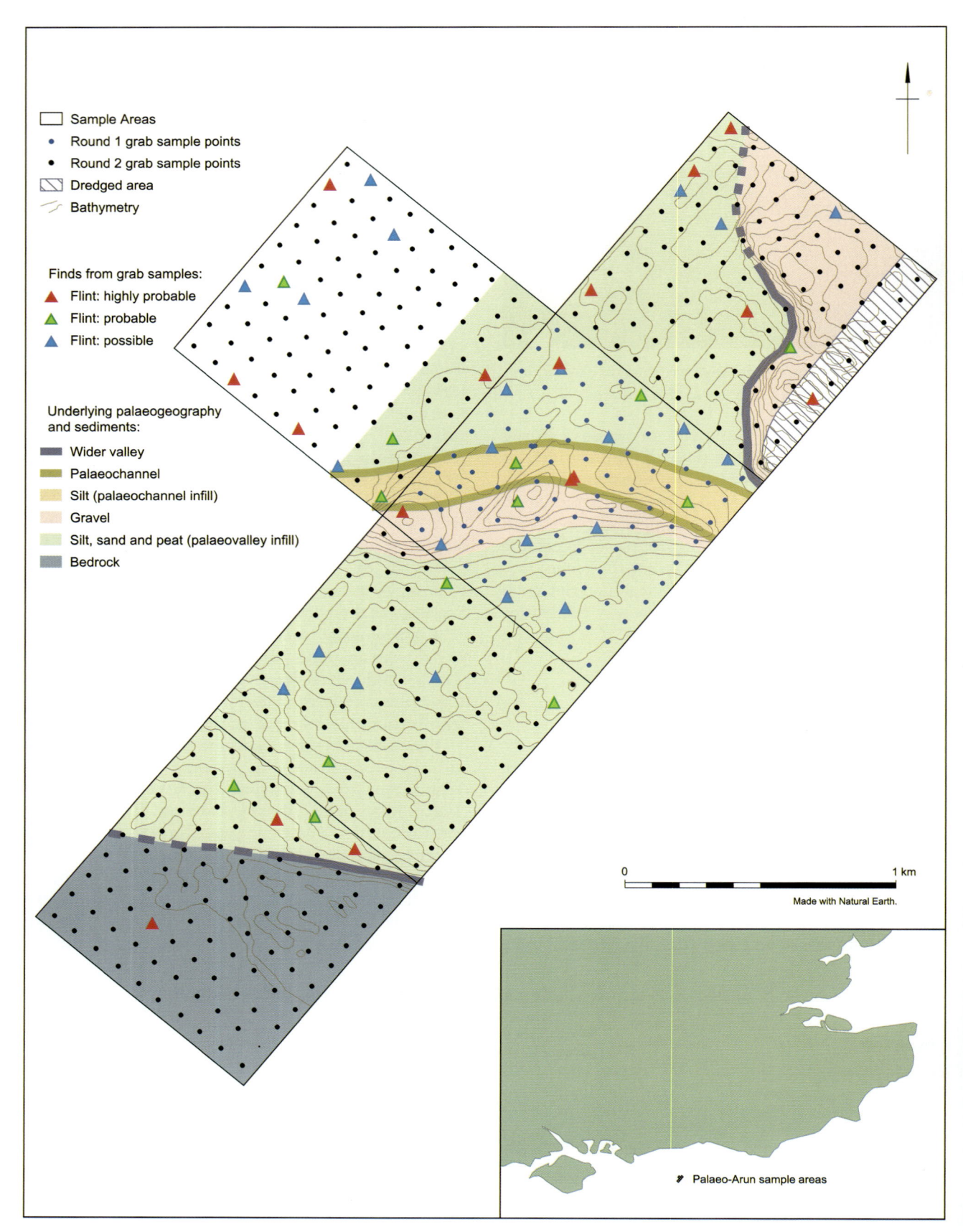

Figure 5.2 Systematic sampling strategy used in the Palaeo-Arun and location of sample stations where worked flint were recovered. Worked flint was primarily recovered from the channel and within the wider valley

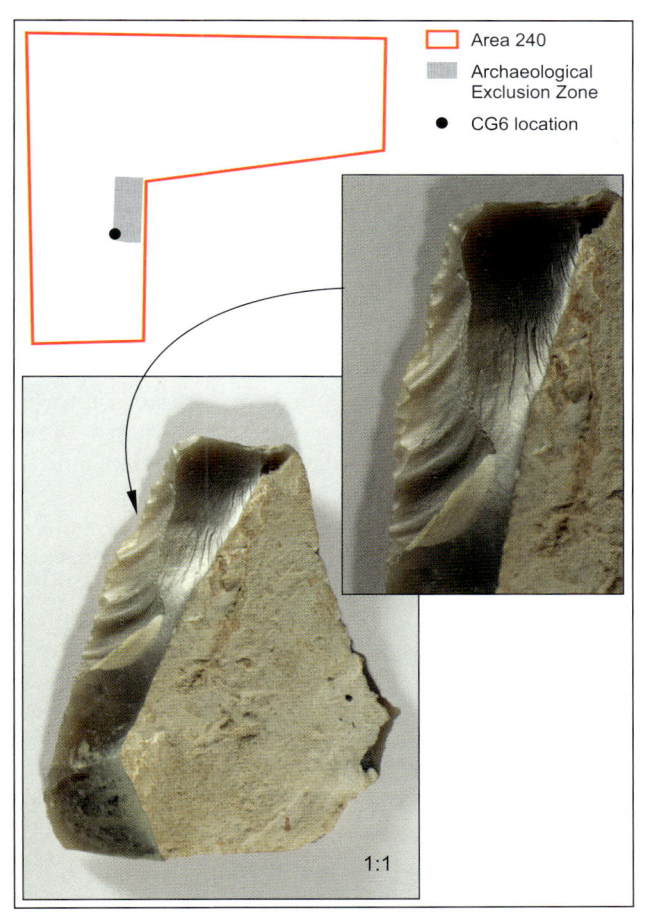

Plate 5.2 Broken secondary flake recovered from clamshell grab sample during the East Coast Regional Environmental Characterisation survey

minor blade-like characteristics or resemblances to secondary flakes, piercers and pseudo-microliths. Examples classed as possible flint artefacts included small flakes and chips that could be anthropogenic in origin, but lacked sufficient indicators to be conclusively diagnosed. These flint finds provided possible evidence of human occupation of the Palaeo-Arun area.

Although it is considered that the flint was arguably of anthropogenic origin, in many cases it is difficult to distinguish from flint produced by mechanical or natural marine processes. The assemblage as a whole was dominated by primary chips and flakes with linear and crushed platforms symptomatic of mechanical fracture and there is an absence of larger material such as axes, cores and diagnostic debitage which may be expected within gravel deposits. Although there is some possible doubt as to the anthropogenic nature of the flints found associated with the Arun, the results proved that as a methodology grab sampling had potential to recover possible artefacts. The flint assemblages appeared to be associated with broader palaeogeographic landforms identified in the geophysics data. However, the large sample spacing (100 m) precluded identification of small concentrations of struck flint.

The use of grab sampling as an archaeological evaluation and mitigation tool was discussed in depth at a seminar funded by English Heritage in December 2007. The seminar was organised by Wessex Archaeology and attended by English Heritage, Cefas, archaeological trusts, consultants and academics (Wessex Archaeology 2008g). Sampling strategies and grid densities were discussed with particular emphasis on the practical application of high density grids and applicability of other available grab techniques. Alternative methods to grab sampling such as anchor dredgers, benthic trawling and suction samplers operated by divers were also discussed, along with the potential for using a combination of techniques. The seminar concluded that more work would be required in order or evaluate grab sampling as a methodology. It was acknowledged that grab sampling has the capability of recovering archaeological material, but that a larger sample size and a finer sample interval may increase the potential of the technique.

Hydraulic clamshell grab samples are routinely recovered by the aggregate industry for resource assessment purposes to identify sediment and particle size analysis. Sample sizes vary but are generally 300 litres sediment in which some stratigraphy is preserved, typically up to 0.5 m. The principal advantage of the clamshell grab over the Hamon grab is the volume of sediment recovered. There are very few instances of clamshell grabs being used to sample for archaeology. Fedje and Josenhans (2000) detail the use of a clamshell grab to target specific, high potential archaeological sites identified from the modelling of geophysical survey data off the coast of British Columbia. There, a single piece of worked flint was successfully recovered from delta floodplain deposit in water depths of approximately 50 m.

Also, in June 2009, as part of the multi-disciplinary East Coast REC project a clamshell grab (sample station CG6) specifically targeting the Area 240 Archaeological Exclusion Zone (AEZ) recovered a broken secondary flake (Limpenny *et al.* 2011). The surviving dimensions of the piece are approximately 60 x 43 x 9 mm, although a transverse break means that the piece was originally considerably longer. This is comparable in size and condition to the artefacts discovered in 2007/2008. Although deliberate retouch is absent, both lateral margins have been used. The right margin has light edge damage towards the distal end; the proximal two-thirds, however, show evidence of more robust use. The left margin is almost entirely cortical, but one short section comes to a cortex-free point, which appears to have been used as a piercing tool. The butt is facetted, and the platform edge has been prepared. One facet on the dorsal surface has a light patina; otherwise the piece is unpatinated and in very good condition, showing no signs of rolling, staining or damage congruent with its having been redeposited or having undergone any disturbance subsequent to its original loss/discard (Pl. 5.2). The recovery of worked flint, albeit two single pieces in each of two surveys, further indicated the potential for clamshell grabs to recover archaeological material.

The discovery of archaeological material in Area 240 afforded the opportunity to further test seabed sampling techniques in an area where archaeological material was recovered and was considered likely that some material should remain.

Area 240 Seabed Sampling Survey

Between 27 July and 4 August 2009 Wessex Archaeology, with support from EMU Limited, undertook a sampling survey on the multicat vessel *Gray Mammoth*, although three of these days were lost due to poor weather. The principle sampling methodology was the clamshell grab. Additionally, a two metre scientific beam trawl was trialled. This equipment had never been used specifically for archaeological work and this project afforded the opportunity to trial the system. In order to aid the choice of sampling locations, video was also acquired (Wessex Archaeology 2010b).

Sampling Strategy

Previous use of grab samples for archaeological purposes has focused on a regular sample pattern (100 x 100 m) which was a suitable sampling strategy for areas of seabed with potential but with no known archaeology. However, due to the environmental conditions in Area 240 a regular sampling pattern was not suitable. The sediments targeted for archaeological material are overlain by a unit of recent marine sediments which form a veneer or as sandwave features, generally up to 4 m high. As such, the sampling needed to target the troughs of the sandwaves where the overburden was at its thinnest thereby targeting the older pre-transgressive sediments. This meant towing equipment cross-current which was not ideal, but it was the only way to target the sediments from which the archaeological material was originally dredged.

Prior to the survey a number of transects were selected within the 3.5 x 1 km survey area of the ARZ which were oriented in the sandwave troughs and orientated east–west (Fig. 5.3). The transects varied in length and, wherever possible, targeted different sediment units based on the geophysical interpretation. Three transects were surveyed during the survey (T1, T2 and T3).

For each transect a visual inspection was carried out followed by a trawl transect. Finally, based on the results of the visual and trawl assessments a series of clamshell grab samples were acquired. Each technique is discussed in more detail below.

Positioning

Spatial positioning of the sampling equipment was key to the study, particularly the grab samples. Accurate positioning of the equipment, by use of an acoustic tracking system enabled specific features identified in the geophysical data to be targeted, provided actual position of acquired samples and, as such, allowed good comparison to the geophysical data. Also, good accuracy of samples allowed for repeat visits to the sample site.

Vessel positioning was provided by Differential Global Positioning System (DGPS) with positional accuracy of better than one metre, which updated every second. Positioning of the video sledge, beam trawl and clamshell grab was provided by the Ultra Short Baseline (USBL) SCOUT acoustic tracking system from Sonardyne. In a USBL system, the position is calculated by measuring range and bearing from the vessel mounted transceiver to the submerged transponder, which emits acoustic signals.

Video/Stills Photographs

Two video systems were trialled. The first was a towed sledge system where the camera was attached to a sledge and towed along the seabed behind the vessel providing a continuous video image of the seabed. The second system was ELVIS, a bespoke low visibility camera designed by EMU Limited and acts as a drop-down video system with two variable frequency Halogen lighting units (Pl. 5.3). This allowed the acquisition of seabed images under poor visibility conditions. In both cases a Kongsberg OE14–208 new generation digital stills colour camera video system was used. All images were recorded topside on a Panasonic 400 GB HDD incorporating a Digital Versatile Disc (DVD) recorder. The towed video system was initially used as this would have the advantage of providing a continuous image of the seabed conditions along the proposed transect. Although weather and sea state conditions were good during the survey, the strong cross-currents meant that the video system was very difficult to control and visibility was poor. Low-visibility was caused by the natural environmental conditions coupled with the mobile sediments being kicked up by the sledge.

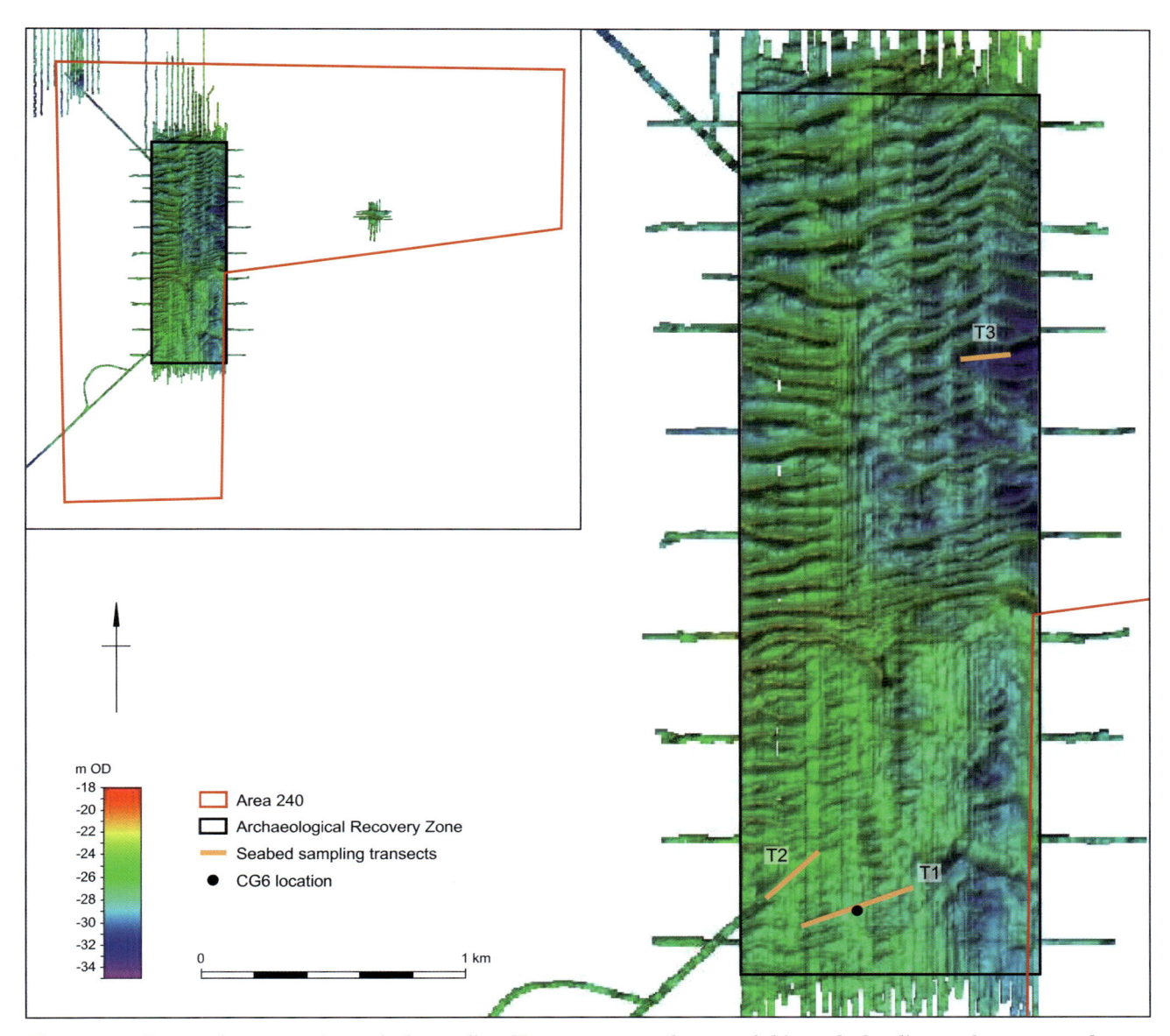

Figure 5.3 Proposed transects for seabed sampling. Transects targeted areas of thin seabed sediments between sandwaves identified on the bathymetry data

Plate 5.3 Deployment of the ELVIS drop-down camera

The use of the ELVIS low-visibility drop-down camera was a success. Although this method did not provide a continuous image of the seabed, the system provided a series of clear images at set locations along the transects. At each site the camera housing is lifted approximately one metre above the seabed and then dropped again onto the seabed. On contact with the seabed there is a one or two second delay whilst the sediment clears before a photograph is taken. This delay also ensured a good, steady positional reading from the transponder beacon. Up to eight photographs were taken at each of the locations, within a 10-metre radius. This allowed a general appreciation of the seabed sediments within the area immediately surrounding the proposed location, rather than multiple photographs of the same area of seabed. Photographs were acquired at six locations

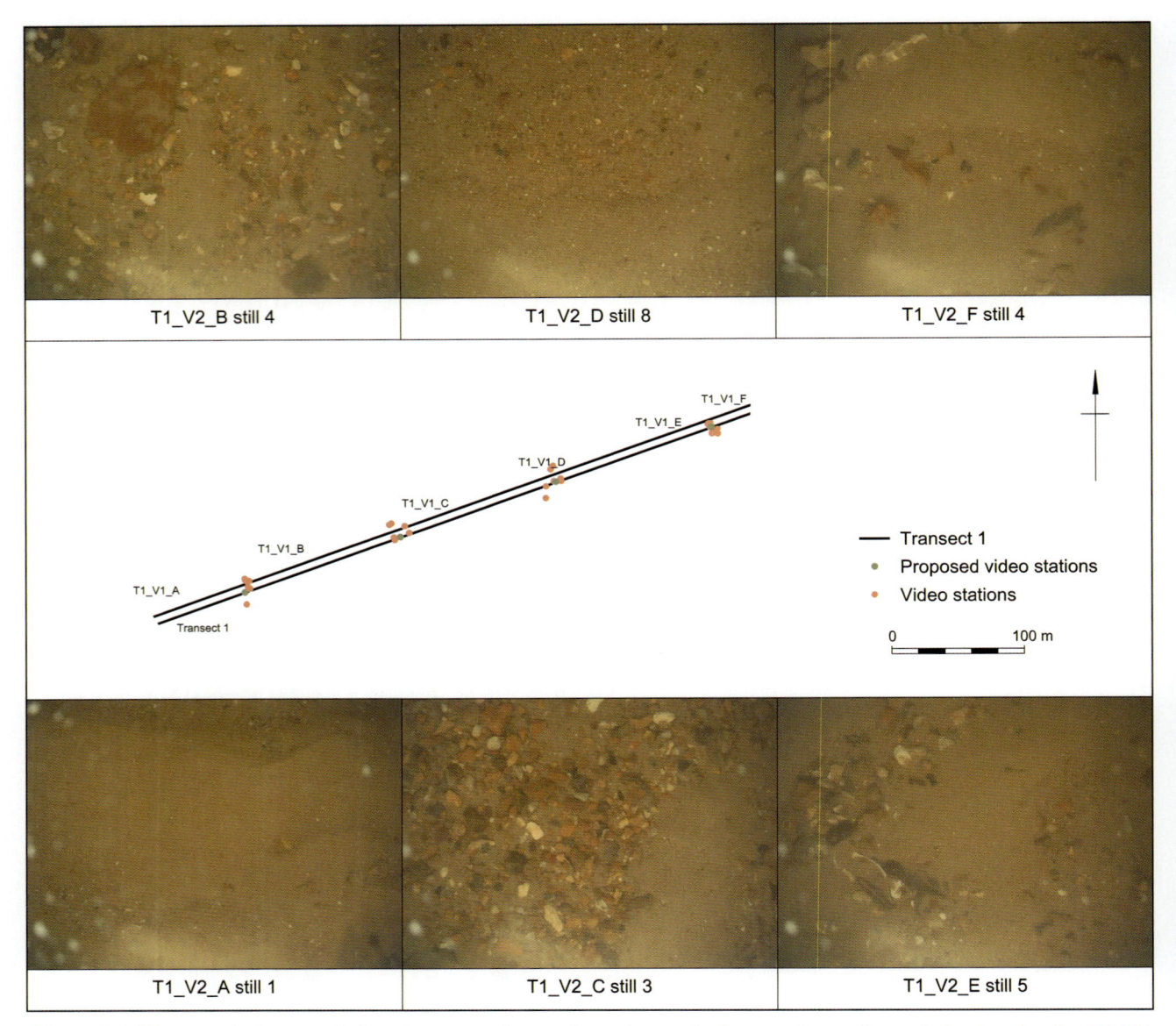

Figure 5.4 Transect 1 photograph locations at each sample station and photographs on the seabed taken by the ELVIS drop-down camera. Photographs indicate a generally sandy seabed with patches of gravel and the occasional cobble

at approximately 100 m intervals on Transect 1 (Fig. 5.4), six locations at 50 m intervals for Transect 2 and five locations at 50 m intervals along Transect 3. Although not as effective as a continuous-towed system, the low-visibility drop-down camera produced a series of usable images that allowed an assessment of the general nature of the seabed were used to select the initial grab sample locations. Additional grab samples were then placed along the transect based on the quality of the initial grab results.

Two metre Scientific Trawl

A number of faunal remains and Palaeolithic artefacts have been reported from commercial trawling activity over the years (Godwin and Godwin 1933; Coles 1998; Glimmerveen *et al.* 2004) and the recovery of prehistoric artefacts, including hand axes, by shell fishing in the Solent (Wessex Archaeology 2003) supported the trialling of benthic beam trawls for archaeological application.

For the purpose of this project, due to the apparent proximity of the artefacts to the seabed, a 2 m scientific trawl was selected for trial. The 2 m beam trawl is used for both habitat mapping and biological monitoring surveys and is the recommended device for the sampling of epifauna at marine aggregate extraction sites (Cefas 2008).

A 2 m scientific beam trawl with chains with 5 mm cod end mesh was used during the survey. The trawl was deployed from a crane on the bow of the vessel and was towed on a pair of bridles attached to a single warp line (Pl. 5.4). The trawl was towed

Plate 5.4 Recovery of the 2 m scientific trawl and sample from Transect 1. The sample comprises gravel, reworked peat and wood as well as seabed fauna

approximately 90 m behind the vessel, approximately three times the average water depth. A transponder beacon was attached to the warp wire and USBL positions were recorded. The beam trawl was towed along the transects but staying online was difficult due to the strong seabed currents.

On recovery of the beam trawl the contents of the trawl were emptied in a deck tray, photographed and visually inspected. The sample was then transferred to a sampling table and where appropriate was washed through a 10 mm sieve and inspected for any archaeological material. The trawl had never been used for archaeological inspection before and had mixed results. The main disadvantage of using the beam trawl is the lack of accurate positioning. Although a beacon was attached to the trawl and the position of the trawl was known at all times whilst on-line, if any artefacts were recovered their position would only be known to lie on the transect, not at any specific point.

Grab Sample Acquisition and Processing

A 280 litre Kinshofer hydraulic clamshell grab was used to recover samples at the Area 240 site. The grab had a maximum penetration of 0.5 m, depending on the coarseness of the sediment.

Sample locations were initially selected based on the results of the photographic survey. The sample interval was also increased based on the on-going results of the grab sampling. A total of 30 grab

Plate 5.5 Recovery of the 280 litre hydraulic clamshell grab sample and on board processing of the sample through a 10 mm sieve

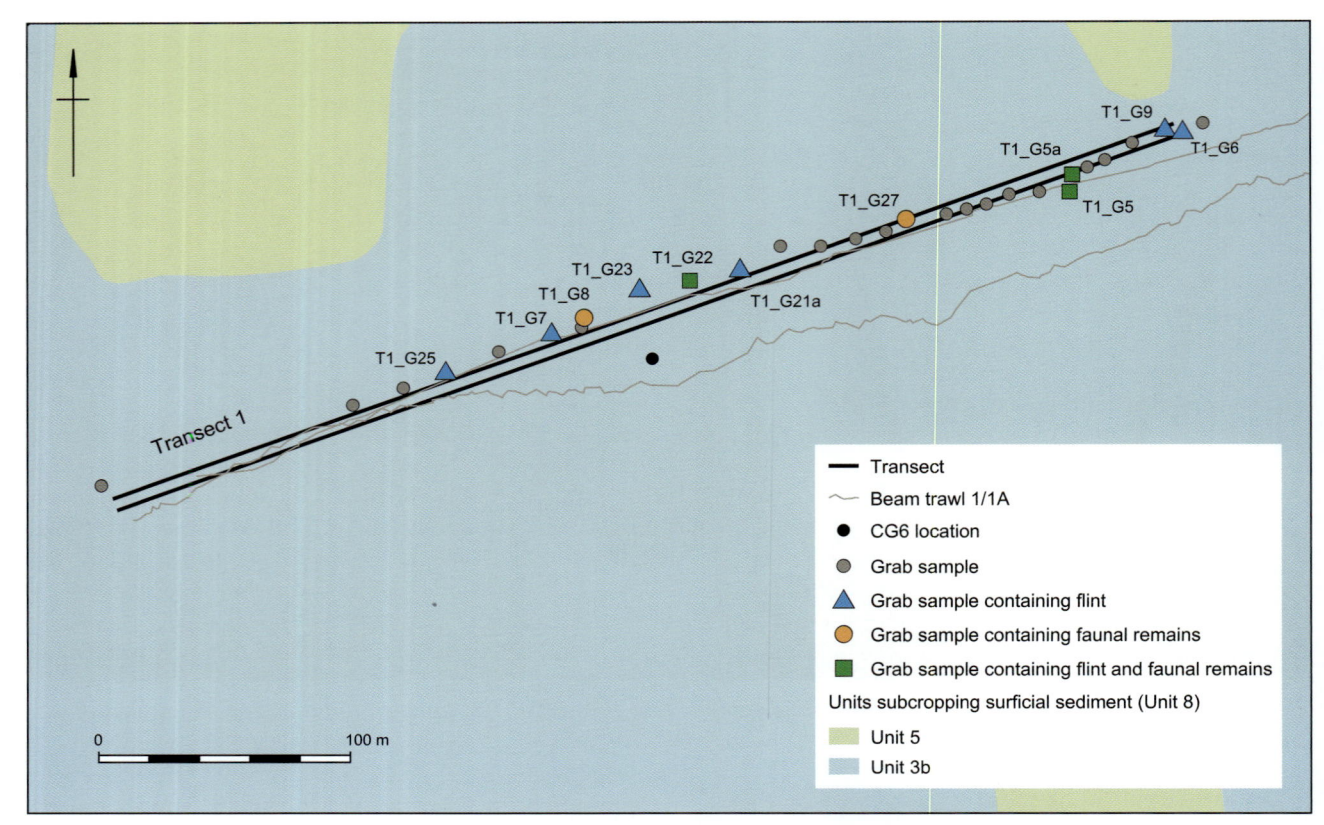

Figure 5.5 Sample locations along Transect 1 targeting the floodplain deposits (Unit 3b) from which flint artefacts and faunal remains were recovered through clamshell grab sampling

samples at 28 locations were acquired on Transect 1, eight samples at seven locations along Transect 2 and at five sample locations along Transect 3. The high number of samples along Transect 1 is indicative of the perceived potential for archaeological material as the samples were acquired, as discussed in detail below.

On recovery, the clamshell grab was lowered to the deck, the top doors of the bucket were opened and the sediment was photographed. The main grab bucket was then opened slightly revealing a thin profile of sediment which was recorded and photographed before the contents of the grab buckets were emptied onto a deck tray. The bulk sample was photographed and visually inspected. Any obvious archaeological material was removed, bagged and labelled (Pl. 5.5). A 40 litre whole-earth bulk sample was taken and stored in four 10 litre tubs. These were labelled and stored on the vessel to be processed post-survey. The remaining sample was transferred to a sampling table and washed through a 10 mm sieve and inspected for archaeological material. The 40-litre bulk samples were wet sieved through a nest of sieves including 10 mm and 4 mm. The less than 4 mm residue was quickly scanned and discarded. The greater than 4mm and 10 mm residues were sorted wet and any finds kept for analysis.

Transect 1

Transect 1 was situated in the south-west of the ARZ within the area of dredging associated with the discovery of the original archaeological material. The transect was 500 m in length and targeted the Saalian floodplain sediments (Unit 3b).

The photographs generally indicated a gravelly, shelly, sandy seabed. Gravels were more prominent in troughs and generally comprised sub-rounded to sub-angular flint and occasional pebbles and cobbles were noted. The seabed was noticeably more gravelly to the north of the transect although it was noted to be patchy. Based on the results of the video survey the transect for the beam trawl was moved 5 m north to target the more obvious gravelly sediments. However, due to the currents it was not possible to follow the transect. Two trawls were completed. The first sample was smaller than desired and so the heavier trawl with tickler chain was added. The second recovered sample contained seabed fauna, gravel, waterlogged wood and some rolled peat. The size of the rolled peat and wood fragments indicated that adding more weight and a tickler chain increased the size of the particles retained in the sample. However, as this was not trawled along the same line as the previous trawl is not necessarily a comparable sample. The rolled

Plate 5.6 Flint flake recovered during clamshell grab sampling (T1_G5). Location of sample is provided in Figure 5.5

Plate 5.7 Flint flake recovered during clamshell grab sampling (T1_G23). Location of sample is provided in Figure 5.5

nature of the peat and wood indicated this was reworked from terrestrial sediments, probably the early Holocene terrestrial deposits (Unit 7).

The grab sampling survey along Transect 1 was successful and archaeological material was recovered. In total, 13 flint flakes and seven pieces of bone were recovered from the clamshell grab samples along Transect 1. The recovered flint consisted entirely of flint flakes, the by-products of flint tool manufacture. Although it can be difficult to distinguish flint flakes of anthropogenic origin from those that occur naturally, due to post-deposition effects, of the 13 flakes recovered all are of probable anthropogenic origin, but eight are more obviously genuine artefacts (Table 5.2 and Fig. 5.5). Two of the flakes, from samples T1_G22 and T1_G25, are broken mid-sections of tertiary flakes with well-defined flake scars and are characteristic of hand axe thinning flakes found at other sites. The flake from T1_G5 shows evidence of platform preparation (Pl. 5.6) and the flake from T1_G23 is a primary flake of clear anthropogenic origin with a striking platform and point of percussion (Pl. 5.7). The flakes from T1_G5a and T1_G21a show evidence of deliberate, systematic debitage. Two flakes, from T1_G6 and T1_G9, are hard-hammer struck and although less convincing than the other flint flakes, are probably of human workmanship. The remaining five flint pieces (G7, G9, G25 comprising three pieces) are considered possible artefacts but open to doubt and may have been formed by natural processes. There are no diagnostic features which allow the flakes to be dated based on typology. The flakes predominantly showed post-depositional surface modifications or were heavily weathered suggesting recovery from an eroding surface or secondary context. In this sense the flakes are comparable to those of the original discovery.

Additionally, nine pieces of bone were recovered from the clamshell grab samples along Transect 1. Most of the pieces recovered were unidentifiable. Two pieces of fossilised bone were recovered from T1_G5,

Table 5.2: Flint recovered from seabed sampling

Grab station	Sediment description	Flake description	Discrimination
T1_G22	10YR 7/3 Very pale brown fine, medium and coarse (predominantly coarse) grained sand. Occasional small flint rounded to angular up to 90 mm diameter. Occasional small broken molluscs. Occasional iron oxide mud concretions	Mid-section of a tertiary flake, with well-defined conchoidal rings on the ventral surface. The dorsal surface also has a number of converging negative flake scars. It has a slightly dipping profile. These features, including the way in which it has broken, have been noted on other handaxe thinning flakes	Most likely
T1_G25	10YR 5/4 Yellowish brown gravelly sand. Gravel is predominantly flint, rounded to angular up to 100 mm diameter. Occasional quartzite up to 25 mm diameter. Occasional quartz up to 15 mm diameter	Flake is similar to that from sample T1_G22. This flake also lacks the proximal and distal ends, so valuable details of the technology are lost. However, the dorsal surface has a number of residual flake scars, which form a radial pattern	Most likely
T1_G5	10YR 6/2 Light brownish grey gravelly sand. Sand is fine to medium grained. Occasional sub-rounded quartz and quartzite up to 40 mm diameter. Occasional iron oxide concreted mudstone up to 50 mm diameter	A very thin flake in mint condition and unstained. The point of percussion is located at the edge of the flake. It is possible that this flake was removed by natural processes, however there are apparent traces of platform preparation and other facets suggest that this is a product of debitage	Most likely
T1_G23	2.5Y 4/1 Dark grey silty sand mottled light grey. Sand is coarse. 4% Black ?degraded organic stain mottling. Occasional small to medium rounded to angular flint to 75 mm diameter. Occasional broken molluscs including *Ostrea edulis*. Occasional sub-rounded quartzite up to 70 mm diameter. One rounded granite pebble 10 mm diameter. One brown conglomerate 40 x 4 mm	Stained and patinated primary, hard-hammer struck flake. The most convincing feature that indicates human production is the clear striking platform and well positioned point of percussion well back from the edge of the core	Most likely
		Three small flakes, all open to some doubt	Possible
T1_G5a	10YR 4/2 Dark greyish brown sand. Sand is medium grained. Moderate broken shell. Gravel is predominantly flint, sub-rounded to sub-angular up to 100mm diameter. 5% Quartz, rounded to sub-rounded up to 30 mm diameter. Occasional brown sandstone (up to 4 mm diameter). Occasional iron oxide mudstone lumps up to 150 mm diameter	A heavily rolled flake with a glossy finish. It is naturally backed. The proximal end is missing, having been chipped by recent damage; however the presence of clear conchoidal rings on the ventral surface and similar well defined traces on the dorsal surface, indicating a previous removal, suggest that this flake is genuine	Probable
T1_G21a	10 YR 6/3 Pale brown sand. Sand is medium to coarse (predominantly coarse) grained. Frequent small to medium up to 80 mm diameter rounded to angular flint. Very occasional quartz up to 35 mm diameter. Occasional ?limestone up to 45 mm diameter. Two ?metamorphic pebbles. Moderate broken shell up to 3 mm diameter	Elongated hard hammer struck flake. It is unstained and unpatinated. The presence of other flake scars suggests that it is product of deliberate, systematic debitage	Probable
T1_G6	10YR 4/3 Brown sand. Sand is fine to medium grained. Occasional quartz sub-rounded up to 30 mm diameter. Very occasional rose quartz. Occasional rounded to angular flint up to 70 mm diameter	Primary flake that is both patinated and stained. It is hard hammer struck. The striking platform is plain and the point of percussion is well positioned on the striking platform and not a glancing blow	Probable
T1_G9	10YR 5/3 Pale brown sand. Sand is medium to coarse grained. Frequent small, sub-rounded to sub-angular up to 60 mm diameter. Occasional quartz up to 22 mm diameter	Clearly hard-hammer struck and is part of a 'compound' removal, where a flake was removed with this one at the same time and the same blow. While not certain, it is possibly due to human workmanship	Probable
		Small flint is principally cortical and not convincing	Possible
T1_G7	10YR 5/4 Yellowish brown gravelly sand. Flint up to 110 mm diameter rounded to angular (predominantly angular). Occasional quartz and quartzite up to 43 mm diameter. Occasional broken shell	A small patinated and rolled primary flake, open to some doubt as to its formation	Possible

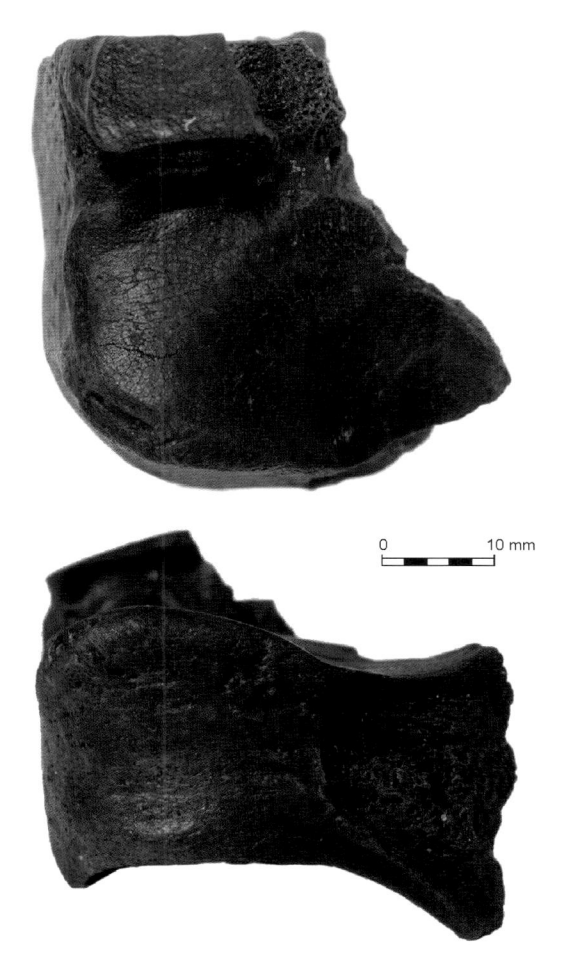

Plate 5.8 Cervid/bovine centrotarsus recovered during clamshell grab sampling (T1_G5). Location of sample is provided in Figure 5.5

one unidentifiable and one broken bovine or cervid centrotarsal (Pl. 5.8). Two other pieces of unidentified fossilised bone were recovered from T1_G5a. A small piece of unidentifiable bone was recovered from the sample T1_G8 and two pieces of large, probably terrestrial mammal bone were recovered from T1_G27. A fish vertebra, probable salmonid, was recovered from T1_G22. The occurrence of terrestrial mammal bone in T1_G5 and T1_G27 from the north-eastern end of the Transect 1 is of interest given that the original discovery included terrestrial mammal bone.

Transect 2

Transect 2 was situated approximately 200 m to the north-west of Transect 1 and was approximately 270 m in length (Fig. 5.3). The transect was confined by the presence of sandwaves to the west and east of the line. Transect 2 targeted floodplain sand and gravel deposits (Unit 3b) in the east and finer-grained sands (Unit 5) in the east.

The photographs indicate a seabed comprising predominantly shelly sand with occasional gravel and cobbles. The dominance of shelly sand indicated a thicker unit of marine sediments (Unit 8) than observed on Transect 1. A beam trawl survey was conducted along the line. Again, due to the cross currents in the area it was difficult to tow along the proposed transect. The recovered sample from the beam trawl contained seabed fauna, gravel, waterlogged wood and some rolled peat.

Initially six grabs were recovered, one at each video station. Although the photographic evidence indicated that sands dominated the seabed surface, and as such presented a low potential for flint artefacts, grabs were acquired to ground truth the locations and to establish whether the sand formed a substantial thickness or a thin sand layer overlying gravels. The grab samples predominantly comprised shelly sand with occasional gravel. Two flint flakes were recovered although these may be the product of gravel abrasion. No faunal remains were recovered.

Transect 3

Transect 3 was situated in the north, approximately 2 km to the north of Transect 1 and targeted older sediments of Yarmouth Roads Formation (Unit 2) (Fig. 5.3). The majority of the photographs indicated a seabed comprising shelly sand with occasional gravel and rolled peat observed, as well as thin sheets of iron-red sediment. These sheets generally appear to be cobble-sized but only a few millimetres thick. These features are not *in situ* and some are observed sticking out of the seabed. On recovery of the clamshell grab samples; these were interpreted as iron oxide (FeO) sheets. The beam trawl sample comprised seabed fauna, small gravel, waterlogged wood and some rolled peat. The majority of the non-faunal sample comprised waterlogged wood.

The grabs were mainly composed of silty sand with occasional gravel and pebbles. Iron oxide sheets were observed in each of the grabs and were generally small pieces, up to 50 mm across although larger sheets were observed in the centre and west of the transect. No flint artefacts were recovered from samples and the only faunal remains recovered was a possible dolphin vertebra.

Discussion

Sampling for archaeological material using established techniques proved successful. The video stills were useful in ascertaining the nature of the seabed sediments, the homogeneity of the sediments along the length of the transect and aided in the initial selection of the grab sampling positions. Although a towed system is considered more useful than a drop-down camera system in that it provides a continuous view of the seabed rather than images at individual points, it is not feasible in areas where the visibility is poor and, as such, the use of a drop-down camera is preferable.

Forty-eight grab samples recovered approximately 8250 litres of sediment or 16.7 tonnes, and numerous flint flakes and bones were recovered (discussed further in Chapter 6). Although only a very small proportion of Area 240 has been sampled along three small transects, the systematic targeted approach of video and grab sampling has shown that flint artefacts and palaeoenvironmental samples can be recovered using this methodology. A further advantage of the clamshell grab is the volume of the sample recovered compared to previous trials using the Hamon grab, as well as deeper penetration and detail of stratification within the upper 0.5 m.

Of the three methods trialled the beam trawling proved least useful, although some wood and peat were recovered, indicating the presence of terrestrial landsurfaces. The environment was not conducive to towed equipment and further trials would be required in more favourable conditions in order to fully evaluate the use of trawling for the recovery of archaeological material. The use of similar established techniques, such as the scallop dredge, could be considered. Scallop dredges, used during fisheries surveys (and, as the name suggests, for scallop dredging), can be deployed from the side of a vessel and towed along a transect comparable to the beam trawl deployment. The scallop dredge consists of a triangular iron frame, with toothed cross bar set at an acute angle to scrape the scallops from the seabed into a bag that is attached to the frame. The belly of the bag is made from steel rings, to withstand chafing, and the top part is of a sufficiently large mesh net to allow gravel dug up with the scallops to pass through. However, the dredge retains finds greater than 50 mm and as such would retain any large pieces of flint such as hand axes. The disadvantage of this type of dredge is that it would not be effective for smaller flakes and as it is an invasive sampling technique, may damage any smaller artefacts.

Based on the results of this trial, trawling should not be dismissed entirely. In collaborative surveys attended by biologists and archaeologists, such as the regional characterisation projects (eg, Limpenny *et al.* 2011), beam trawling was carried out for biological purposes and the recovered sample assessed by an archaeologist. With appropriate training it may be possible for biologists undertaking beam trawl sampling to assess the samples archaeologically. Samples or artefacts could then be retained for study by suitably qualified archaeologists.

Although a series of flakes and bones were obtained by sampling in this area, no hand axes or evidence of a confined archaeological site were found. This is likely to be a reflection of the methodology employed rather than an indication of the presence or absence of material. Despite the choice of sampling location being judgement led and based on as much evidence as was available and the systematic, targeted approach using photographic stills and grab sampling, only a very small proportion of the area was sampled.

One of the clear outcomes from the seabed sampling conducted at Area 240 is that the volume of sediment acquired is an important consideration. Although a significant increase in the volume of sediment was processed in the Area 240 survey compared to past surveys using the Hamon grab and completed more successfully, it was concluded that volume of sediment was still an issue. The original archaeological discovery was recovered from 11 dredge trips during which approximately 55,000 tonnes of sediment would have been dredged from the seabed. Even comparing a single dredge of 5000 tonnes this was approximately 250 times more sediment from which 13 flakes were recovered during the Area 240 grab sampling survey. Arguably, the return of the flakes was also due to the judgement-led sampling strategy. However, volume of sediment is still a major issue.

Dredging for Archaeological, Palaeontological and Palaeoenvironmental Material

Following the conclusion of the Area 240 project in March 2011 Wessex Archaeology continued discussions with Hanson Aggregate Marine Limited and English Heritage regarding the remaining potential for artefacts in Area 240 and potential mitigation that could result in the reduction or removal of the AEZ in the future. English Heritage determined that any changes to the AEZ would be dependent on further assessment of absence or presence of archaeological material. Although the clamshell grab sampling indicated that there was cultural material present, particularly to the south of the archaeological exclusion zone, only a very small area of the seabed was sampled. Questions remained over the extent of the archaeological material, particularly the flint artefacts and it was clear that the volume of sediment sampled remained a limiting factor.

The original discovery was recovered from a large volume of dredged sediment; however there was no control over the position or geological context of the dredging. Trial dredging with control over these parameters afforded the assessment of certain sediment units identified during the geophysical and geotechnical data review and to assess large volumes of sediment.

In 2011 Wessex Archaeology were commissioned by Hanson Aggregate Marine Limited to undertake a programme of archaeological monitoring of aggregate dredging activity on board a dredging vessel and at the receiving wharf at SBV Flushing (Wessex Archaeology 2012a). The aim was to assess potential mitigation strategies, with regards to future long-term aggregate licensing applications, in an area of known archaeology. The project, in consultation with English Heritage, trialled the method of bulk sampling using standard aggregate dredging plant with the aim of recovering artefactual material in industrial processes for the purpose of evaluation. Additionally, the aim was then to use any material to evaluate the presence or absence, distribution, character, quality and preservation of Palaeolithic artefacts within Area 240, and in particular the AEZ.

There were two main elements to the survey which comprised a walkover survey of aggregates within the hopper of the dredger, *Arco Adur*, immediately after dredging the target transects. Secondly, visual examination of material during offloading at the wharf and assessment of the oversize material at the wharf was undertaken.

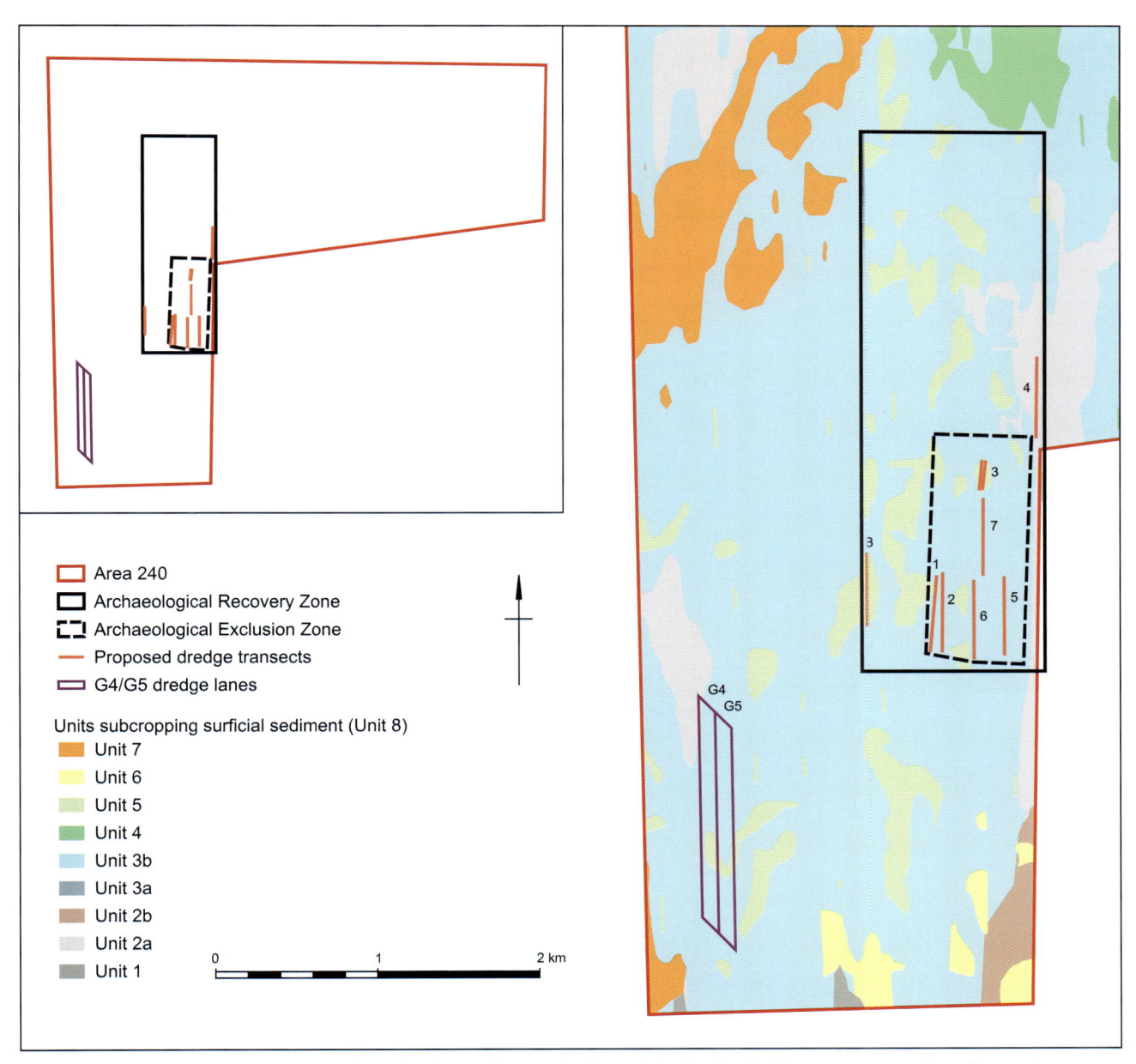

Figure 5.6 Proposed transects and dredging lanes for monitoring of dredge loads. Transects were selected targeting particular sediment units

Sampling Strategy

The vessel *Arco Adur* is a trailing suction dredger similar to most aggregate dredgers operating in British waters and capable of holding 5000 tons of aggregate. The sediments targeted for archaeological material were primarily situated in the AEZ and, as such, repeat dredging of full loads of 5000 tons was not permitted under the sampling licence. Removal of 1000 tons per sampling event was granted which equated to approximately 500 m long transects.

Eight transects of approximately 500 m long were proposed, developed in consultation with English Heritage and Hanson Aggregate Marine Limited; they targeted specific geological units, based on the 2005 and 2009 integrated geological interpretation (Fig. 5.6). The transects (Table 5.3) were selected to target single geological units where it was considered that the chance of recovery of artefacts was high, ie, in areas where there was limited overburden. A variety of sediment units were targeted providing good spatial coverage principally, but not exclusively, within the exclusion zone. The transects were orientated north–south to account for tidal conditions.

In order to reduce impact on commercial activities and comply with the sampling licence (a Marine Licence was required to extract sediments from the transects from within the AEZ) an initial load of 4000 tons was taken from two pre-existing dredging lanes established by Hanson Aggregate Marine Limited within Area 240. Dredging lanes G4 and G5 are situated in the south-west corner of the area approximately 1 km from the existing AEZ (Fig. 5.6). After the initial dredge the load was walked over and any archaeological material identified was recovered

Plate 5.9 An archaeologist assesses dredged sediments for archaeology on board the Arco Adur

(Pl. 5.9) to try to ensure that any material subsequently identified was derived from the subsequent sampling from specified transects. A second load of 1000 tons was then taken from one of eight predefined transects within the ARZ of Area 240. A walkover of the second load was then undertaken.

Dredging vessel: Assessment of Dredge Loads

The walkover on the dredge load surface was generally a success, although there were limitations to the method, not least the health and safety issues; the presence of sand in the cargo led to difficulties in the walkover of the material. The sand slowed the

Table 5.3 Sediment unit targeted on each transect

Transect number	Target sediment unit	Notes
1	Unit 3b	Transect targeted Unit 3b (MIS 8/7; Middle Palaeolithic) in the area where previous worked flints and bone were located. 2009 bathymetry and geophysics data indicate minimal overburden (<1 m)
2	Unit 3b	Transect targeted Unit 3b in the area where previous worked flints were located. Situated 40 m east of proposed Transect 1. Some sandwaves in northern section (less than 2 m high)
3	Unit 5	Split transect in order to target Unit 5 within the exclusion zone. Minimal overburden observed along transects except for a sandwave around 2 m high to the north of the transect
4	Unit 2	Targeted Unit 2 (Lower Palaeolithic). Targeted to confirm the absence of lithics. It is a possible source of older faunal remains. Some small (approximately 2 m) sandwaves in the central and northern section of the transect. Situated outside of AEZ but within the 2009 survey area
5	Unit 3b	Targeted Unit 3b in an area known to have been heavily dredged prior to the implementation of the AEZ to assess presence/absence in this part of the AEZ
6	Unit 3b	Targeted Unit 3b to the east of the area where flints and faunal remains were sampled in 2010 (Wessex Archaeology 2010b) to assess presence/absence in this portion of the AEZ. Veneer (<1 m) recent sediments throughout the transect
7	Unit 3b	Targeted Unit 3b in the central area of the AEZ to assess presence/absence in this part of the zone. Veneer (<1 m) recent sediments throughout the transect
8	Unit 5	Targeted unit 5 outside the AEZ but within the 2009 survey area to assess presence/absence in this unit. Recent sediments are generally <1 m thick throughout the transect
G4/G5	Predominantly Unit 3b	Predominantly Unit 3b sediments with small area of underlying Unit 2 and small area of Unit 5

Plate 5.10 An archaeologist ready to assesses oversize material at the wharf

draining of the cargo, required to make it safe to walk over, and also masked much of the other material. As the aggregate is conveyed into the hopper, a certain degree of sorting of material occurs and larger heavier elements may be drawn down into body of the cargo. Items of archaeological interest, such as hand axes and mineralised (partly fossilised) faunal remains, are relatively large and heavy and it is probable that the sorting of aggregates masked items like these in some cases. Conversely, some lighter material of archaeological interest such as wood and non-mineralised faunal remains was drawn to the surface of the hopper. It was found to be relatively easy to recover lighter materials as they tended to gather in depressions on the surface of the hopper formed during draining of the load.

Material of archaeological interest, including flint artefacts, was recovered by the walkover from each of the eight transects and from the initial loads taken within lanes G4/G5 (Fig. 5.6). Finds from the hopper included organic material, such as wood, bone, teeth, struck flint, amber and peat. This material is discussed in more detail below.

SBV Flushing Wharf

The wharf at Flushing is a largely automated plant for the processing of aggregates, covering an area of three hectares. It consists of a wharf where dredging vessels can dock and offload material and a yard with processing machinery and several large stockpiles of unsorted and sorted material.

Due to the mechanism by which the aggregate is removed from the vessel, the loads (from the dredging lanes and the transects) were mixed and provenance of any finds at the wharf could not be stated with certainty. At the wharf oversize (>63 mm) material was inspected at the screening table (Pl. 5.10). It was considered that there is a limited potential for archaeological remains within aggregate grades below 63 mm although smaller artefacts such as debitage could be found, these were deemed less likely to be spotted amongst the aggregate. This proved to be a highly successful approach and a number of worked flints were recovered including two well-preserved hand axes.

Material processed through the de-wooding installation was also assessed for archaeological material. This is a large tank which shakes the load (<63 mm) and flushes water through it from below. Lighter material is carried to the upper part of the tank where it is flushed off to the side. This lighter material is largely comprised of shell fragments with some wood and other organic remains. A single inspection of the de-wooding material inside the plant (thought to represent material from Area 240 over the last six months) recovered a single piece of mammal bone.

The oversize pile was also assessed, although to a limited extent. Unlike many British wharves, oversize material (>63 mm) is not immediately crushed during the sorting process but is output to a temporary oversize stockpile. It was from this oversize stockpile that the initial discovery was made in 2007/2008.

Recovered Flint

Altogether, 153 pieces of lithic material were collected from the dredger (Pl. 5.9) and at the wharfside from the eight loads that made up this project, of which 24 were subsequently positively identified as artefactual material (Table 5.4). Of these, five were securely attributed to material dredged from lanes G4 and G5 and seven from the transects. The remaining 12 were from mixed material, and of these nine were identified from Trip 1, although the greater number recovered is likely attributed to the longer duration spent at the wharf with this load, rather than the suggestion of increased quantity of flint.

The flint artefacts were associated with the transects targeting Unit 3b and it is noticeable that the loads targeting Units 5 and 2a did not yield any lithic material. Unit 3b (*c.* MIS 8/7) is the predominant sedimentary unit in the dredging lanes G4/G5 and it is considered likely that the archaeological material recovered from these dredging lanes is also associated with the unit. Associations between the sediment units and the archaeological material are discussed in Chapter 6.

Three Palaeolithic hand axes were recovered during this work (Fig. 5.7). Two cordiform (heart-shaped) implements (Find nos 1000 and 1011) of

Table 5.4 Flint artefacts recovered during dredge monitoring (artefacts recovered from a single sediment unit (not mixed) are highlighted)

Transect	Find no.	Sediment unit	Source	Flint type	Description
1	00	2, 3b and 5	Mixed sediment load from wharf	Flake	Large, mainly cortical flake, unpatinated, unstained, three points of impact, hard, slightly rolled, 1 inverse removal
1	1000	2, 3b and 5	Mixed sediment load from wharf	Hand axe	Cordiform on flake blank, ventral surface flaked sufficient to thin butt, dorsal covering flaking, lightly stained, sharp (135 x 95 x 39 mm)
1	1002	2, 3b and 5	Mixed sediment load from wharf	Flake	Large tertiary flake, hard hammer, plain butt, lightly stained, partially radial flake scars, possibly from Levallois flake core (95 x 107 x 19 mm)
1	1006	2, 3b and 5	Mixed sediment load from wharf	Flake	Large primary flake, unpatinated/unstained, mint/sharp, could be modern on condition but included due to well-placed point of impact (137 x 106 x 37 mm)
1	1007	2, 3b and 5	Mixed sediment load from wharf	Flake	Large flake, stained, sharp/slightly rolled, some modern edge damage (102 x 103 x 23 mm)
1	1008	2, 3b and 5	Mixed sediment load from wharf	Flake	Large primary flake, thermal dorsal surface, cortical butt, stained, slightly rolled/rolled
1	1009	2, 3b and 5	Mixed sediment load from wharf	Flake	Stained secondary, hard-hammer struck flake, slightly rolled/rolled, cortical butt, clumsy crushed impact (86 x 82 x 23 mm)
1	1011	2, 3b and 5	Mixed sediment load from wharf	Hand axe	Hand axe with plano-convex cross section, probably made on flake. Both sides with covering flaking. Lightly stained, slightly rolled, tip absent (113 x 80 x 23 mm)
1	1012	2, 3b and 5	Mixed sediment load from wharf	Core	Core fragment with a pot lid fracture, but with relict flake scars (two deeply invasive and one alternate) that are rolled suggesting the recently formed pot lid may have come from a humanly modified block
1B	1018	3b	Transect 1	Flake	Rolled secondary flake, damaged distal end, rolled, stained, distal part broken
				Flake	Tertiary flake, cortical butt, lightly rolled/rolled, lightly patinated
2	1038	2, 3b and 5	Mixed sediment load from wharf (G4/G5 and Transect 2)	Flake	Large tertiary flake, stained, slightly rolled/rolled, plain butt, uncertain mode, from flake core (77 x 114 x 55 mm)
2	1039	2, 3b and 5	Mixed sediment load from wharf (G4/G5 and Transect 2)	Flake	Large primary hard-hammer struck flake, rolled stained, plain butt (97 x 112 x 21 mm)
2A	1024	2, 3b and 5	G4/G5	Flake	Broken thinning/shaping flake, opposed scars, linear butt
2B	1025	3b	Transect 2	Flake	Possible flake
4A	1045	2, 3b and 5	Mixed sediment load from wharf (G4/G5 and Transect 4)	Flake	Large hard-hammer struck secondary flake. Possibly represents a stage of hand axe roughing out/shaping. Three unidirectional flake scars. Good quality flint, unstained, slightly rolled, unpatinated. Plain butt, no preparation
5	1085	2, 3b and 5	Mixed oversize pile.	Hand axe	Ovate/cordiform. Tip absent, well executed bifacial covering flaking, lightly stained, sharp (87 x 92 x 23 mm)
5A	1054	2, 3b and 5	Mixed sediment load from wharf (G4/G5 and Transect 5)	Flake	Tertiary flake, slightly rolled, lightly stained, no preparation, possible signs of soft percussion
5B	1058	3b	Transect 5	Flake	Broken hard-hammer struck secondary flake, light differential staining, sharp. Unidirectional flaking, plain butt (68 x 57 x 22 mm)
7A	1087	2, 3b and 5	Mixed sediment load from wharf (G4/G5 and Transect 7)	Flake	Flake stained sharp, opposing dorsal scar patterns
				Flake	Flake stained sharp clear butt, hinged distal end
7B	1088	3b	Transect 7	Flake	Lightly stained flake, butt unclear, transverse dorsal scars may be anthropogenic
				Flake	Rolled flake with parallel flaking scars lightly patinated. Possibly represents hand axe thinning
8B	1096	3b and 5	Transect 8	Flake	Faceted butt, sharp, lightly patinated, hard, distal tip absent but almost certainly blade, possibly retouched

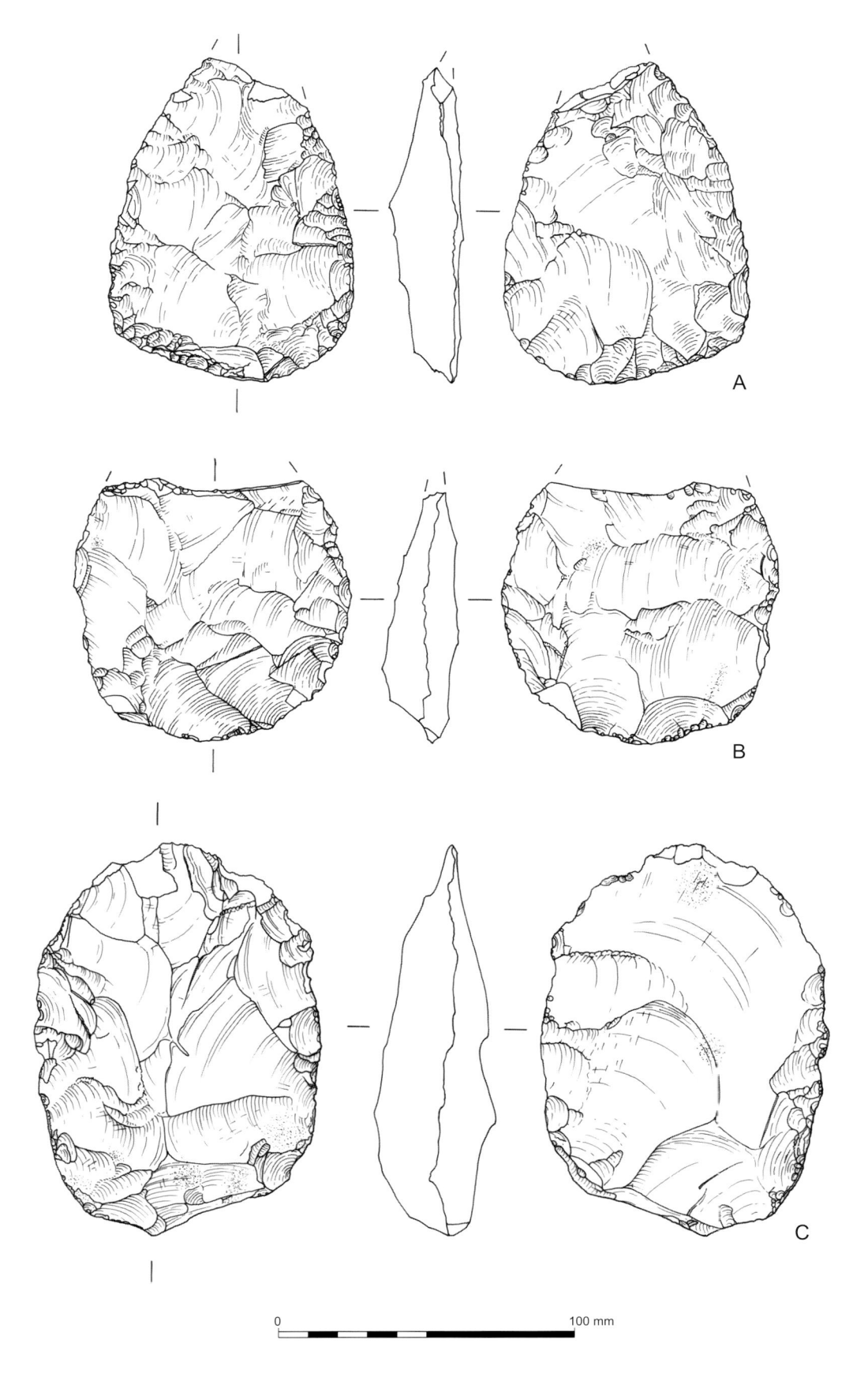

Figure 5.7 Illustrations of the hand axes recovered during monitoring of a dredge load. A) Find no. 1011; B) Find no. 1085; C) Find no. 1000

Wymer's (1968) Type J were recovered from Trip 1 and a cordiform/ovate (SF 1085 – Type J/K) from Trip 5. Both hand axes from Trip 1 were characterised by plano-convex cross sections, suggesting that they may have been made on flake blanks. A sufficiently large area of the ventral surface of the blank survives on Find no. 1000 to confirm this method of manufacture. All implements are lightly stained, slightly rolled, and in a sharp condition. Flaking patterns indicate a process of thinning using bifacial, covering or invasive flaking to create the desired form. No cortex survives on any of the implements.

Twenty-one flakes were recovered and accounted for 88% of the Palaeolithic artefacts. They were dominated by relatively large pieces, generally 80–90 mm, but occasionally over 100 mm long. This undoubtedly relates in part to collection from the 'oversize' reject heap and conveyor, a collection strategy that does not favour recovery of smaller pieces. In addition, some components of debitage were undoubtedly winnowed out by fluvial activity when the gravels were laid down or reworked. Nevertheless, the production of such large flakes suggests that the locality in which they were found contained plentiful supplies of large nodules of flint for tool manufacture.

Most of the flakes are undiagnostic and, with the exception of two possible hand axe thinning flakes (Find no. 1024 from Trip 2A and Find no. 1088 from Trip 7B), may have derived from either core or core tool manufacture.

The assessment of material from the original collection in Area 240 (see Chapter 2) indicated that Levallois (prepared core) technology was represented among the collection. No Levallois cores were recovered from the most recent phase of work; however a large tertiary flake (Find no. 1002) and probable broken blade (Find no. 1096), which has a carefully faceted butt, seem likely to have been removed from a prepared core, *sensu* Bordes (1961). This small addition to the total material collected from the seabed appears to confirm a Levallois element to the core reduction strategy employed in Area 240. The presence of this material is in keeping with the previous conclusions of work on material from the site.

Most of the Palaeolithic artefacts are in a sharp to slightly rolled condition, with only occasional pieces that had been more heavily rolled. This included a recently broken fragment from Trip 1 (Find no. 1012) listed as a core fragment that also displayed two deeply invasive flake scars that were heavily water-worn. These flake scars are considered more likely to have been removed as a result of a deliberate Palaeolithic flaking strategy, rather than by marine activity. This implies that the majority of material has not been subject to significant movement and can be considered to be from a (near-) primary context prior to dredging. The condition of individual artefacts indicates that most of the worked flint assemblage is from a disturbed context, but is unlikely to have moved far from its original point of discard. The fresh unabraded component exists with isolated, more heavily rolled material that may represent older artefacts that have been more extensively reworked in the gravel.

The consistency of the artefact composition, especially hand axe form and condition, are immediately comparable to those pieces recovered from Area 240. These factors support the argument that the additional artefacts, both hand axes and flakes, were derived from the same geographical location and sediment deposits as those discoveries made in 2007 and 2008.

Recovered Palaeontological and Environmental Material

Faunal remains were recovered from almost every load but were noticeably more prevalent within the AEZ. A total of 36 faunal pieces were recovered and included a number of mammoth teeth and large mammal bones including a metatarsal. One tooth was identified as a *Mammuthus primigenius* upper molar (Andy Currant pers. comm.). Much of the other material was indeterminate but thought to be from large herbivores. No specialist analyses of these faunal remains have been undertaken to date.

Approximately 220 pieces of waterlogged (and sometimes mineralised) wood were recovered during the dredger walkover and post-dredging sorting at the wharf (Table 5.5). On analysis the majority of the wood was fresh and little mineralised suggesting a Devensian or Holocene age. Although these were associated with the transect targeting the older Cromerian sediments (Unit 2) but is likely that these were recovered from the recent seabed sediments (Unit 8), having been eroded from their original context. More mineralised wood fragments were recovered from Saalian and Devensian sediments (Unit 3b and Unit 5) and the antiquity of the wood does not discount the possibility that the wood is contemporary to these sediments. It cannot be absolutely stated that the remains originate from a Saalian gravel or indeed a hitherto unidentified organic (?interstadial) lens within the gravels, they may relate to reworked Cromerian material, but there is a strong possibility they are contemporary. No further analyses concerning the identification of the remains to genus or, where possible, species, was carried out as part of the project. Further analyses could provide information on the local environment and vegetation of the time and allow further correlation between the wood and the sediment units.

Table 5.5 Recovery of palaeontological and environmental material recovered during monitoring of dredge loads

Transect	Unit	Wood	Faunal remains	Peat
1	3b	—	-	Peat (250 ml) not in secure context. Possibly early Holocene in origin
2	3b	One fragment rounded highly mineralised (black, heavy, iron-rich, haematite?). Potentially pre-Devensian	2 pieces	—
3	5	35 heavily mineralised orange (iron-rich) fragments. Potentially pre- late Devensian. (Reminiscent of >50 ka material looked at in previous phase)	1 piece	—
4	2a	Approximately 20 fragments mainly fresh, angular, some mineralised. Plus 28 pieces, good fresh condition, a few part-rounded. Likely Devensian/ Holocene from condition	—	—
5	3b	Eight fragments, three rounded, rest angular. Possibly Devensian/ Holocene from condition One large mineralised roundwood fragment. Potentially pre-Devensian 32 pieces fresh but rounded wood (seabed surface material?) with ten pieces jet/shale. Probably late Devensian/ Holocene from condition 29 fragments fresh but some rounded. Two possibly worked pieces: one curved probably worked and polished on inner edge (resembles a handle or eg, skate), one thin rounded piece with possible facet on tip	3 pieces	—
6	3b	17 fragments including twigs and mature wood. Some part rounded but excellent fresh condition. Some possibly recent coniferous pieces. One small rounded piece removed as possibly worked. Probably late Devensian/Holocene from condition	21 pieces	—
7	3b	Twenty-eight pieces including twig and root material as well as mature. In relatively good condition but several partly mineralised (some orange, Fe), could possibly be contemporary with targeted bed	5 pieces	—
8	5	Eight small pieces, fresh but rolled/semi-rounded. With three pieces jet/ shale, one piece coke/slag. Possibly late Devensian/Holocene from condition, also reworked	—	—
G4/G5 or mixed load	Mainly 3b, some 5 and small area of 2	Nine very large pieces of mature and large branch wood. Relatively good condition but several are mineralised (dark, possibly haematite). One piece sub-rounded. From condition possibly Devensian or earlier Seven wood fragments recovered from hopper sample associated with Transect 1. Good fresh condition, including roundwood pieces. Possibly Devensian/ Holocene from condition One large fragment mature (oak) wood and one soft rounded wood fragment	4 pieces	Mineral-rich peat without a secure context. Potentially mixed sample (80 ml)

The wood assemblage also included three fragments which may show possible signs of anthropogenic alteration, although all show evidence of abraded surfaces. These are considered to be *ex situ*.

Two specimens of peat were recovered and are both likely to be reworked rather than *in situ*, similar to the rolled peat observed in the photograph survey and beam trawls during the seabed sampling phase of the Area 240 project.

The trials conducted on the vessel and at the wharf were the first of their kind. The methods were limited in some aspects, in particular with regards to the mixing of sediments which resulted in doubt of provenance in some cases. However, the recovery of archaeological material during this trial has furthered the methodological understanding or assessing archaeological material in a way of working with the industry. Future monitoring of the wharfside processing should focus on the assessment of >40 mm fraction material post-screening and prior to crushing. This would recover the greatest proportion of material of archaeological interest with the greatest control. This operation is also the easiest to adopt at other wharves. There also needs to be greater positional control over the target loads to minimise the occurrence of mixed loads.

Based on the results of the Palaeo-Yare catchment area (detailed in Chapter 4) and the recovery of artefacts from Area 240, work continues in the East Coast dredging block.

In 2012 further trials of this method were successfully undertaken at the Frindsbury Wharf in the UK (Wessex Archaeology 2012b) and a method of operational sampling has been adopted by licensees of the East Coast dredging block at a number of wharves in order to allow the development of a regional framework which would result in a better understanding of the prehistoric archaeological resource in the region in terms of its distribution, significance and the mitigation of effects from dredging.

Summary

A variety of remote sampling methods have been trialled in Area 240. All were established techniques used within the marine industry and all have proved to be successful for sampling archaeological material, to a greater or lesser degree. There is no doubt that sampling archaeology in submerged offshore sites is difficult. The lack of visual assessment and the remote nature of the site require innovative use of available techniques. Furthermore, site-specific conditions need to be accounted for in sampling design. Appreciation of environmental conditions and tight control over location and geological context is imperative in order to assess the context of any archaeological material recovered. Not all techniques and strategies will be applicable to all areas of the seabed and each site needs to be assessed on its own merits.

Since the initial discovery in 2007/2008 a further 37 flint artefacts and a large amount of faunal remains and palaeoenvironmental material has been recovered by clamshell grabbing and trail dredging. Five worked flints were securely attributed to material dredged from lanes G4 and G5 in the south-west of the licence area and 20 from planned transects. The remainder were from mixed load material. The programme of monitoring undertaken during the project has conclusively demonstrated that Palaeolithic material occurs both within and outside of the current AEZ, that the main source of this material is though most probably Unit 3b.

The clamshell grab sampling indicated the presence of material in the southern part of the AEZ and the monitoring of the dredged material confirmed the presence of material within the exclusion zone and also to the west of the current zone. Therefore, this process has confirmed that archaeological material, and in particular flint material, is more widespread than initially indicated. The implication of this, and the Area 240 assemblage as a whole, is discussed in more detail in Chapter 6.

Chapter 6
Examination of the Archaeology, Methodological Approach and Management of Area 240 and Further Afield

Introduction

The work carried out between 2009 and 2012 produced numerous and diverse datasets – geophysical, geotechnical, palaeoenvironmental and archaeological – each subject to its own specialist disciplines of collection and analysis. Combined, the data have led to a comprehensive reconstruction of the development and preservation of the landscapes of Area 240 and have furthered our knowledge on the quantity and extent of archaeological, palaeoenvironmental and palaeontological material. The results of the work carried out are presented, framed by the aims and objectives defined at the project inception. The discussions focus on the recovered archaeological material and its geographical and cultural setting, and an evaluation of the methods used and management and mitigation strategies.

The Area 240 Assemblage Formation and Post-depositional Modification

In order to assess the relationship between the archaeological material and the geology it is necessary to understand the assemblage components, the processes of assemblage formation and post-depositional modification. These processes influence the formation of both primary and secondary context sites. Hosfield (2007, 7) posed a series of questions concerning the interpretative difficulties with assessing the Lower Palaeolithic resource in river systems (both terrestrial and the now-submerged extents). These questions can be adapted to Area 240 and the Palaeo-Yare system, as follows:

1) With which sediment unit, or units, is the archaeological material associated?
2) How old are the sediment units?
3) Over how long a period were the sediments deposited?
4) Does the archaeological material represent a chronologically homogenous sample (ie, is it the product of a 'single' behavioural episode) or a time-average archaeological palimpsest?
5) Is the archaeological material the same age as the sediments, or older?
6) Did the archaeological material originally accumulate in the landscape (through hominin tool discard) at its place of discovery, or has it been displaced from its original context and redeposited downstream?

These questions can be addressed by assessing the technological components of the assemblage, the probable geological context, taphonomic considerations and the likely source of the lithic material, as discussed below.

The Artefact Assemblage

A total of 126 pieces of worked flint were recovered from Area 240 during seabed sampling and dredge monitoring activities (Table 6.1). Recovery locations indicate that the material is not confined to a small, isolated zone of Area 240 but is more widespread. Within the Archaeological Recovery Zone (ARZ) the material seems to be clustered to the south, within the southern half of the Archaeological Exclusion Zone (AEZ). Further material was recovered from dredging lanes in the south-west of the Area.

The hand axes, irrespective of the timing or method of recovery are of similar typology and preservation condition. The large number of Levallois lithics in the assemblage indicate that a substantial component of all lithic material is of MIS 9 age or younger.

In addition to this lithic material, further flint artefacts, with no contextual information, have been reported from Area 240 through the *Protocol for Reporting Finds of Archaeological Interest*. In 2007–2008 two flints, along with two mammoth teeth, were recovered from the aggregate reject pile at SBV Flushing Wharf (Hanson_0180). One of these flints showed possible signs of working and may have been

Table 6.1 Flint artefacts recovered from Area 240

Finds	Hand axes	Cores	Flakes	Total
Initial discovery	33	8 (3 Levallois)	47 (20 Levallois)	88
East Coast REC	—	—	1	1
Seabed Prehistory: Seabed Sampling	—	—	13	13
Dredge and wharf monitoring	3	1	20 (1 Levallois)	24
Total	36	9	81	126

the waste product during the knapping of a flint tool such as a hand axe. Although the positional accuracy is poor their occurrence indicates the potential for further material within Area 240.

Geological Context of the Artefacts

The palaeogeographic reconstruction has revealed a complex history of deposition and erosion within the Area, the interpretation of which has been further complicated by dredging over recent decades. A series of sediment units dating from *c.* 700 ka to inundation at *c.* 8 ka were interpreted, although not as a complete sequence. Within the timeframe of known human occupation of Britain, sediment units were dated from the Cromerian Complex (Unit 2), Saalian (Unit 3a and b), early Devensian (Unit 4), mid-Devensian (Unit 6) and early Holocene (Unit 7).

Prior to this work and work carried out concurrently as part of the East Coast Regional Environmental Characterisation (REC) project there had been very little dating evidence acquired in this region of the North Sea. Saalian (MIS 8–7) and mid-Devensian (MIS 3) deposits had not been previously identified or mapped in the offshore region. Recent work undertaken in the Outer Thames has dated channel deposits to the late Saalian (MIS 6) (Dix and Sturt 2011), indicating that this age of sediment might be more widespread in the offshore region associated with certain channel deposits. Although the presence of early Holocene peat (Unit 7) was known to exist in pockets nearshore (Arthurton *et al.* 1994), the full extent of the early Holocene channel of the Palaeo-Yare has been fully realised through this work. The comprehensive palaeogeographic reconstruction (Chapter 4) has allowed the assessment of the geological context of the archaeological material.

Within the two regions from which artefacts have been recovered (the ARZ and the dredge lanes in the south-west of Area 240) there are four sediment units from which the flint artefacts could have been recovered: Units 2a, 3b, 5 and 8.

Unit 2a is interpreted as a fine-grained sand and clay deposited in a cool outer estuarine or shallow marine environment and forms part of the Ur-Frisia delta-top deposits of the southern North Sea. Optically stimulated luminescence (OSL) dating of the uppermost sediments of this unit returned a date of 735±134 ka (MIS 19; GL 10040) as a minimum age. The typology of the hand axes (Acheulean type) and flakes (including Levallois) post-date the deposition of these sediments. It is unlikely that the artefacts would have been reworked into older sediments and also the fine-grained nature of Unit 2 would have been unlikely to provide a source of raw material for the flint artefacts. Furthermore, although some of the artefacts are in pristine condition the majority show some signs of abrasion, typical of river gravel deposits, of which there is no evidence in Unit 2 sediments. Therefore, Unit 2 can be discounted as the source of the artefactual material.

Unit 3b is an extensive unit observed throughout much of Area 240. The unit comprises a series of sand and gravel layers. Within the ARZ the vibrocores indicate that the base of the unit comprises sand with an increasing gravel component at the top of this unit. The coarser, uppermost sediments are the targets for aggregate extraction and much of the coarser element has been removed through dredging. Within the dredge lanes (G4 and G5) vibrocores indicate a coarser unit. Dating of Unit 3b indicates deposition between 283±56 ka (MIS 9/8) and 207±24 ka (MIS 7). Within the ARZ, just below the effects of dredging, sediment is dated to 243±33 ka (MIS 7). Dating to the accuracy of MIS sub-stages is not possible with the current chronological control and the stratification of the sediment sequence. For example, MIS 7d had similarly glacial conditions as MIS 8 with connotations for human colonisation of Britain (Scott and Ashton 2011). Therefore our synthesis is based upon more general discussions of MIS stages.

Unit 5 is a sand and gravel unit deposited in an estuarine environment and is observed infilling shallow depressions. However, the age of the Unit is unknown. The infilled depressions cut into Unit 3b indicating younger than Saalian age. If contemporaneous with Unit 6 then would suggest mid-Devensian age (MIS 3). However, the Unit could be early Devensian (perhaps between MIS 3/4–5d).

Within the ARZ it is possible that the flint artefacts were dredged from this unit. During the dredge monitoring there was no evidence to suggest that worked flint was recovered solely from Unit 5.

Unit 8 comprises the seabed sediments in the area and contain reworked sediments, sourced, in part, from older depositional units. During the seabed sampling with clamshell grabs Unit 3b was targeted and artefacts were recovered. However, these were heavily abraded and possibly sourced from the uppermost reworked sediments (Unit 8) which would include the underlying Unit 3b sediments. Although associated with a reworked unit, it is unlikely that the flints would have moved far from their source, due to the condition of the finds. No flint artefacts were recovered from Transect 2 or Transect 3. Based on taphonomic considerations, see below, it is unlikely that all the material was recovered from Unit 8.

Within the ARZ, during the recovery of the initial artefacts in 2007 and 2008, it was calculated that 135 km of seabed was dredged in a series of dredge runs. Based on the palaeogeographic interpretation 103 km (77%) targeted Unit 3b compared to 22 km

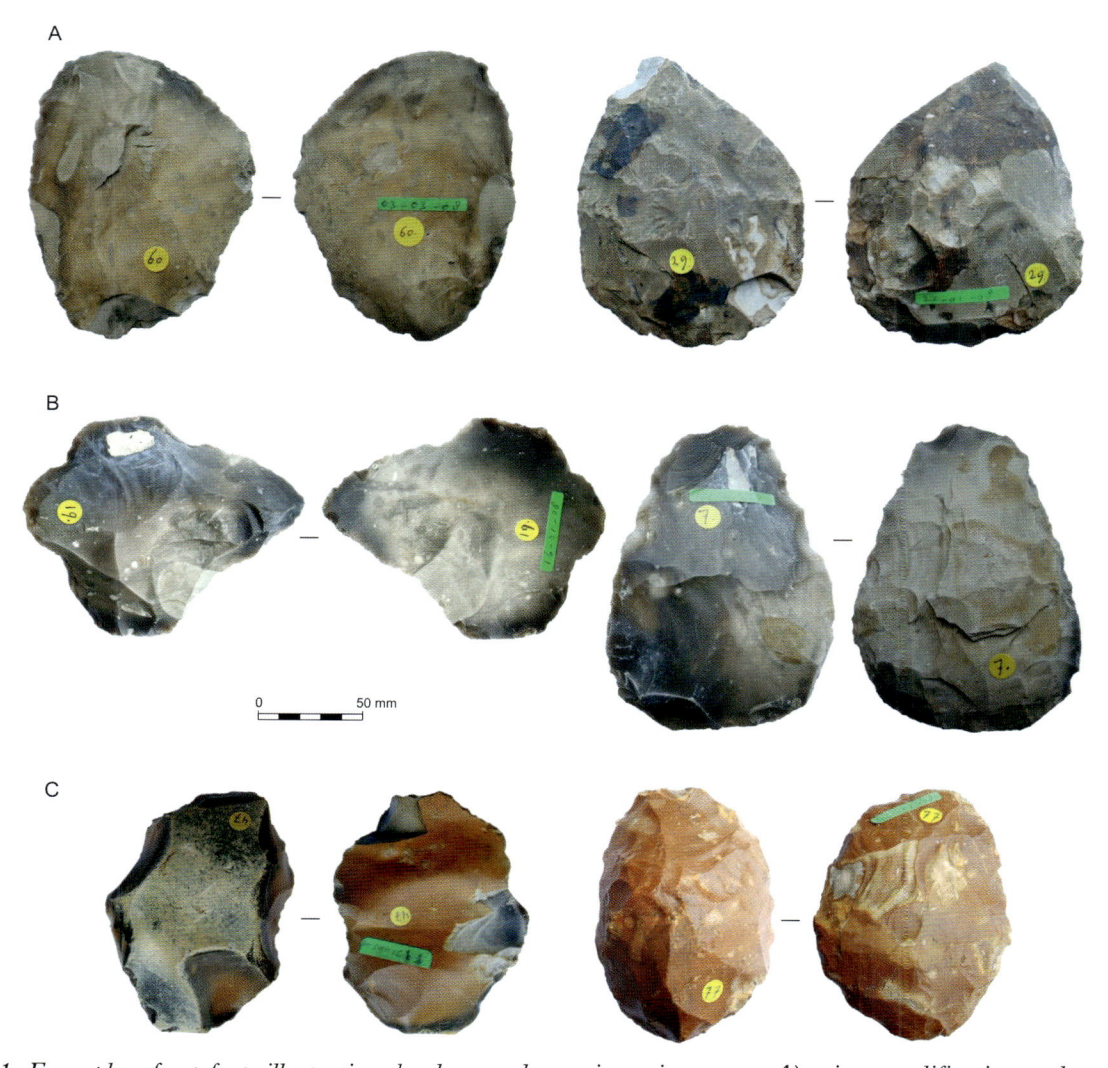

Plate 6.1 Examples of artefacts illustrating the three taphonomic environments. A) minor modifications only; B) post-depositional surface modifications on one side; C) heavily weathered

(16%) of Unit 5 and 10 km (7%) of Unit 2. Although Unit 8 overlies these other units, the draghead of the dredger would be partially buried to target the deeper units and it is unlikely that the material was recovered from Unit 8. It is probable that Unit 2 was dredged in an area where overlying Unit 3b had been removed previously through dredging activities.

During the dredge monitoring in 2011, flint flakes were recovered from transects targeting solely Unit 3b (Transects 2, 5 and 7). Two of the hand axes (Find no. 1000 and 1011) were dredged from dredge Transect 1 but were recovered at the wharf as part of the mixed load from dredge lanes G4 and G5 (Table 5.4). Analysis of the dredge tracks indicates that 13.501 km (81.5%) Unit 3b and only 2.476 km (15%) Unit 5. The third hand axe (SF 1085) was recovered from the mixed load from Transect 5 whereby 11.625 km (85%) of the dredge tracks targeted Unit 3b compared with 1.777 km (13%) targeting Unit 5.

Given the evidence, it is probable that the majority of the flint artefacts were recovered from Unit 3b. However, recovery, at least in part, from Units 5 and 8 cannot be discounted.

Taphonomic Considerations

The condition of the artefacts indicates that several taphonomic processes were responsible for the formation of the assemblage. Three taphonomic environments were identified indicating that the formation and preservation of the assemblage is not straightforward. Plate 6.1 illustrates examples of the flints exhibiting evidence of these environments.

A. Artefacts which show a (minor) colour patina, a light gloss and insignificant edge damage (among these are some flakes and cores, but mainly the hand axes): since the hand axes

show a relatively uniform morphology, a (near) primary context situation is suggested.
B. Artefacts which show predominantly post-depositional, surface modifications (colour patina, gloss and abrasion) on one single side of the artefact (this group mainly consists of flakes): an eroding surface is suggested.
C. Heavily weathered artefacts (heavy colour patina and gloss, and abrasion and battering on both sides of the artefacts), which probably originate from a secondary context such as gravel deposits.

The three different environments do not necessarily indicate that the artefacts are from three different sediment units. For example, the complexity of sediment composition within Unit 3b could preserve *in situ* artefacts in the finer-grained sediments, provide an eroding surface and river gravel deposits (upper coarse-grained sediments).

Comparisons between the Area 240 assemblage of hand axes and a large proportion of Levallois flakes and cores, environmental and taphonomic complexity is observed at other sites in Northwest Europe. This is summarised by Scott and Ashton (2011, 94):

A number of northwest European sites are attributed to cold, open environmental conditions at the beginning and end of MIS 8. The oldest and best-dated site from the region is that of Mesvin IV, which was excavated from two channels forming part of a terrace of the Haine, near Mons. The artefact assemblage as a whole has been rolled to some degree, comprising cores, flakes and handaxes, as well as simple prepared Levallois cores. The latter are very similar to those from Purfleet ... and have been described as 'reduced Levallois' in character ... Some large, fresh, classic Levallois flakes are also present, the condition of which contrasts notably with the handaxes from the site, suggesting a degree of fluvial admixture.

A Local Source of Material and Production?

Area 240 is a source of flint aggregate in modern times, and presumably also in the past when the area was aerially exposed between interglacials. Through examination of the oversize pile at Flushing Wharf, the range of oversize flint pebbles and larger nodules recovered from Area 240 deposits is broad and certainly nodules recovered during aggregate extraction are of appropriate size for tool production, be that by Levallois techniques or Acheulean tradition. The grade of material only accounts for a small proportion of the sediment volume; the vast majority of each dredged load is of finer-grained sediments comprising mainly sands and gravels. Based on the aggregate target in the Area these larger nodules are likely to be sourced from Unit 3b. As proposed in Chapter 2 the source of the flint may have been local which is consistent with these sedimentological observations at the wharf. Research has shown that during the Middle Palaeolithic there are consistent patterns in the distances raw materials have been moved from sources to sites, with locally available material (within 5 km) typically accounting for well over 50% of lithics at a site, regionally-available stone (5–20 km) around 2–20%, and stone from distant sources (30–80 km) less than 5% (Wragg-Sykes 2009).

Due to the sorting processes at the aggregates wharf it is not possible to fully account for smaller elements that may, or may not, have been part of the lithic assemblage at Area 240. The oversize pile only contains >63 mm material and long, thin material is underrepresented. Observations and material collected from the dredged loads prior to sorting at the wharf indicates blades and lithic material smaller than hand axes are present within the Area. However it is not possible to contextualise this material or to assess their relationship to other elements of the lithic assemblage. More difficult, but not impossible, was to recover the broken blade, found within the final dredged load, having been identified by the ship's highly observant Bosun.

The effect of industrial sorting has reduced the behavioural inferences that can be made from the assemblage regarding the production and discard of the lithics. Some reworking of elements of the assemblage is clear but in general, it appears that most of the hand axes and flakes, including the Levallois flakes and cores, have not moved far from where they were discarded.

Site Formation Scenarios

Within the archaeological record of Britain and Northwest Europe hand axe technology is an enduring component of lithic technology with various forms from Lower Palaeolithic and Early and later Middle Palaeolithic periods. Levallois artefacts appear in the British record during MIS 9. Several sites in Northwest Europe region also display a mix of hand axes and Levallois technology (eg, Scott and Ashton 2011) making dating by typology complex and relatively poorly constrained, spatially and temporally. However, three scenarios were presented in Chapter 2 based on the typology of the artefacts. These are revisited below taking into account the geological context and taphonomic considerations (Fig. 6.1).

A. Scenario 1

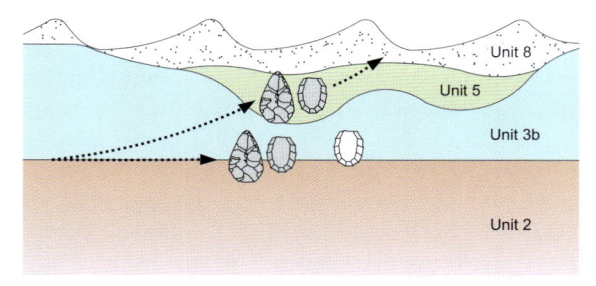

B. Scenario 2

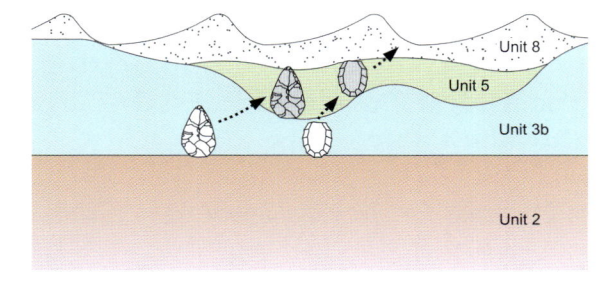

C. Scenario 3

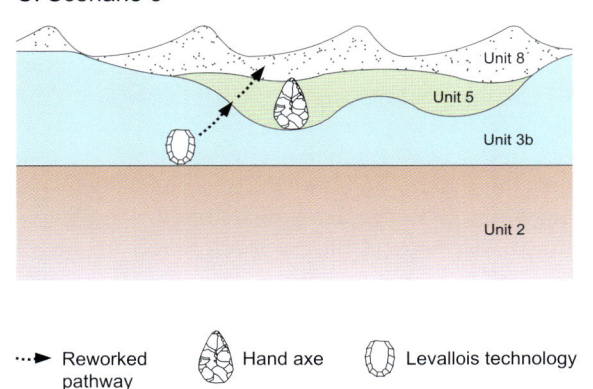

Figure 6.1 Schematic illustrating the three site formation scenarios. Grey symbols indicate reworked artefacts

Scenario 1 argues that the hand axe assemblage represents remnants of an Acheulean findspot dating to pre-MIS 9 (*c.* 500 to 300 ka). However, the sediments from which the hand axes were dredged (Unit 3b or 5) post-date MIS 9. The majority of the hand axes are in pristine condition (taphonomic environment 1) and are thought to be from a primary (or near-primary) context prior to dredging. This reduces the likelihood of pre-MIS 9 hand axes reworked into the younger sediment units (eg, Units 3b, 5 or 8). However, some of the hand axes show evidence of post-depositional weathering (taphonomic environment 2) and could represent pre-MIS 9 artefacts reworked into younger sediments. These reworked hand axes would then be mixed with the younger, Levallois elements of the assemblage, associated with the Unit 3b deposits.

Scenario 2 suggests that the Acheulean hand axes and Levallois products are contemporaneous in geological terms, as it is not possible to resolve which of the multiple sediment layers grouped into Unit 3b are the sources of the various artefacts are, ie, two archaeological periods are conflated into one geological unit dating to around 300 ka or younger (with the Levallois core technologies only appearing in Britain around this time). This is supported by the homogenous character of the used flint throughout the assemblage and also the post-depositional condition on a range of artefacts (both hand axes and Levallois flakes). The dated age and character of Unit 3b sediments also support this scenario. Unit 3b is an appropriate age for the source material for the artefacts and would provide the range of environments (preservation and eroding surfaces) within the sediment deposit. In this scenario it is proposed that the artefacts were all recovered from Unit 3b, although it is possible that some of the abraded, reworked artefacts in this scenario have been recovered from Unit 5.

Scenario 3 suggests that the hand axes are of Mousterian of Acheulean Tradition (MTA) and therefore the assemblage is composed of MIS 8–6 Levallois tools and younger MIS 5d–3 MTA hand axes. This would suggest that the artefacts were recovered from *in situ* contexts from both Unit 3b (Levallois) and Unit 5 (MTA hand axes), indicating two distinct Middle Palaeolithic Neanderthal assemblages. Based on the preservation condition of the flint throughout the assemblage it is difficult to reconcile that the assemblage has been recovered from two separate sediment units, unless a large proportion of Unit 5 is composed of reworked Unit 3b sediments. Also, given the relatively small isolated coverage of Unit 5 and the volume of sediments dredged in recovering the artefacts, it is considered unlikely that Unit 5 has yielded such a volume of artefacts.

On balance and considering the body of evidence that has been gleaned from Area 240 (and the region) scenario 2 is considered the most probable. Several palaeogeographic, sedimentological and taphonomic factors underpin this choice of scenario for the origin of the Area 240 assemblage:

- The sharp, and in some cases, unweathered nature of many of the hand axes (including their relatively homogeneous form and production method) indicates they derive from an *in situ* or near *in situ* context prior to being dredged from the seabed. Therefore it is unlikely that these hand axes have been reworked into a younger deposit (ie, made between 300–500 ka and

eroded into sediments accumulating around 250 ka). However a small number of artefacts are rolled and may have followed this kind of taphonomic pathway;
- Stratigraphic and available chronometric dating indicates Unit 3b is of (MIS 8–7) Saalian age and probably accumulated around 250–200 ka;
- Based on targeted monitoring of dredged loads and proportion of sediments extracted from Area 240, Unit 3b is considered to be the most probable unit from which the flint artefacts (hand axes and Levallois) were recovered. Some of the more reworked elements of the assemblage may have been recovered from Unit 5;
- Within the Northwest European archaeological record, assemblages of hand axes and Levallois technique components have been found to have been included in sediments dated between MIS 8–6 at a number of sites.

Summary

A major theme that is clear from both the palaeogeographic and archaeological record of the Palaeo-Yare catchment as a whole is the intermittent nature of deposition and preservation. This is within the context of the long time periods and regional spatial scales. The records under consideration are a palimpsest of sedimentary units, lithic technology, and floral and faunal remains which have been to varying degrees reworked or written-over by glacial processes, marine transgression and latterly offshore industrial activity. This is not to say that understanding this location or others like it, is fundamentally problematic, but the added complexity of offshore data gathering has required a sustained and involved process of assessment, reassessment and synthesis. A number of significant themes are represented by Area 240: a submerged early prehistoric context; an MIS 8/7 Saalian, probably colder environmental setting; a northerly location considering the age in question; good preservation despite multiple glacial periods and marine transgressions.

The Geographic and Cultural Setting of the Middle Palaeolithic Assemblage within the Palaeo-Yare

The Potential for Archaeological Material in the Aggregate Block

Unit 3b deposits are extensive within the floodplain of the Palaeo-Yare (Fig. 4.6). Although the relationship between the distribution of archaeological material and the overall extent of Unit 3b is not known, it is possible that further flint artefacts are present in other Unit 3b sediments within the aggregate block. Although there is potential it is difficult to state how much and where further artefacts would be found. Given the extent of Unit 3b, it seems unlikely that archaeological material is distributed evenly across Unit 3b deposits. Cultural preferences during the

Table 6.2 Key Middle Pleistocene Palaeolithic sites within the Palaeo-Yare catchment

MIS / Stage	Site	Date (ka)	Technology	Assemblage Type	Environment	References
11 (AAR indicates MIS 11, biostratigraphy indicates MIS 9) Hoxnian	Hoxne, Suffolk	404±33 437±38	Acheulean: ovate & cordate hand axes (Lower Industry); Pointed hand axes, retouched flake (Upper Industry)	Workshop? 3 phases of industry	Warm, Cold; Lacustrine	Singer et al. (1993); Wymer (1999); Penkman et al. (2008); Ashton et al. (2006); Grün and Schwarcz (2000)
?11–9 Hoxnian – Saalian	Whitlingham, Norfolk	(>300?)	Acheulean hand axes, flakes, cores	Workshop	Riverine?	Wymer (1999); Sainty (1927); Pettitt and White (2012)
?<11–7? Hoxnian– Saalian	Keswick Mill Pit, Norfolk	?	Hand axes, flakes, scrapers, (Levallois flakes)	Workshop?	Riverine?	Wymer (1999); Sainty (1933)
Late 8? Saalian	Carrow Road, Norfolk	?	Hand axes/(Levallois flake)		Cool	Wymer (1999); Sainty (1933)
Late 8/early 7 Saalian	Area 240	250–200	Hand axes/Levallois flakes & cores		Riverine, estuarine, cold	This volume

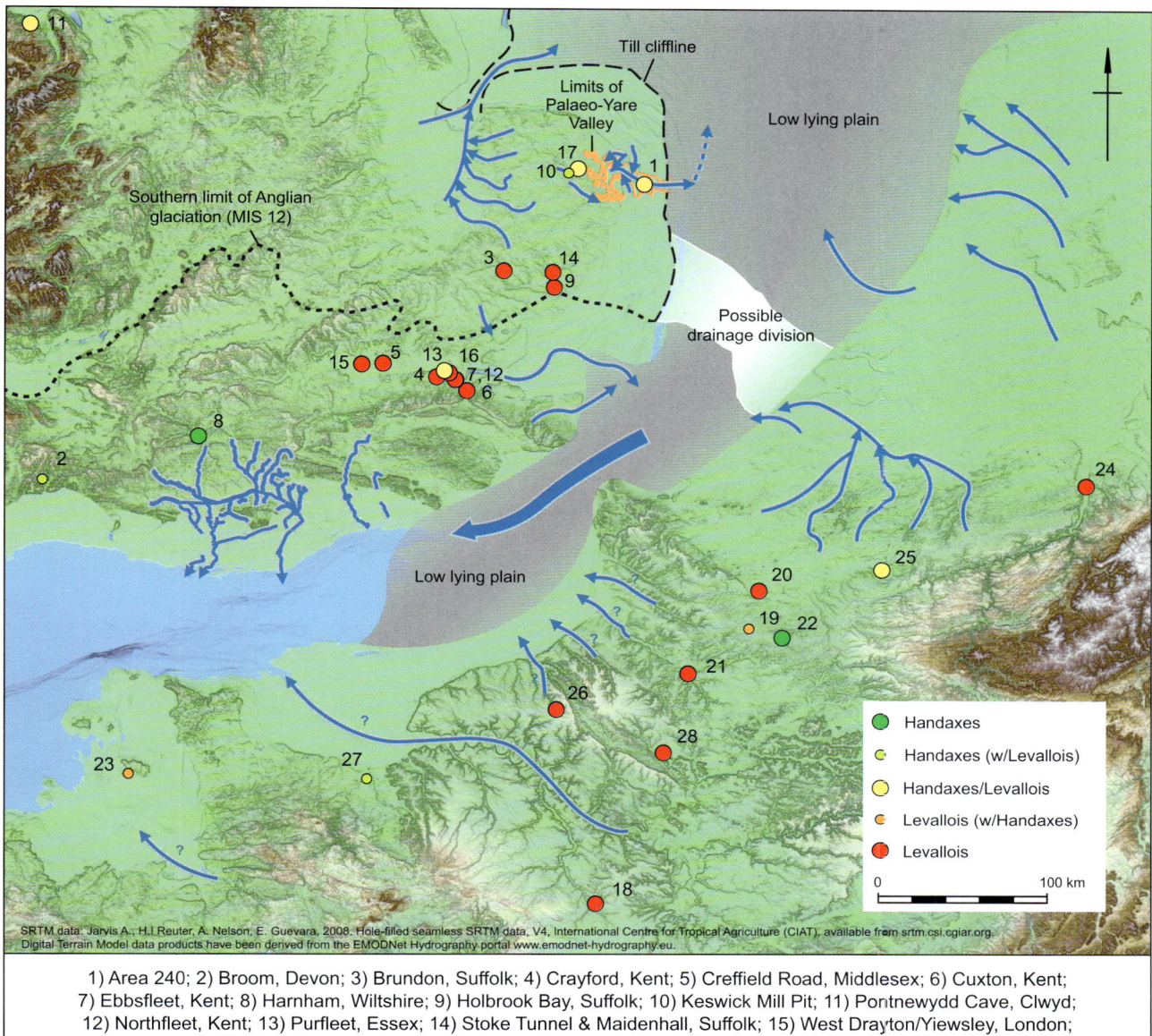

Figure 6.2 Middle Palaeolithic context of the Area 240 assemblage in Northwest Europe c. MIS 9–7 (between around 300–200 ka). The range of sites preserve lithic assemblages dominated by hand axes or Levallois techniques or instances of both in varying proportions. The generalised lithic technology is based on data from Sainty (1933); Wymer (1999); Hosfield and Chambers (2004); Wenban-Smith (2004); Santonja and Villa (2006); Schreve et al. (2006); White et al. (2006); Penkman et al. (2008); Scott and Ashton (2011) and sources discussed in Chapter 2

Palaeolithic, in conjunction with geological process, are likely to be responsible for the distribution of the archaeological material. It may be that the area to the south-west of the channel in Area 240 was preferable to the banks to the north or south-east, though it is difficult to determine site-scale occupation models (based on desirability) and whether there is potential to encounter undiscovered sites elsewhere within the catchment area. It seems likely that the Middle Palaeolithic assemblage recovered from Area 240 represents discrete hotspots of archaeological material within the overall extent of Unit 3b.

Flint finds reported through the *Protocol for Reporting Finds of Archaeological Interest* tentatively suggest that there is potential for the presence of archaeological material within the aggregate block. A piece of worked flint was recovered from Area 360 along with faunal remains, mineralised wood and peat. It was not possible to identify the faunal remains to species but they were large mammals, probably herbivores. Although the materials were reported together there is no evidence of direct association other than being recovered in the same dredge load (CEMEX_0039). Worked flint was also recovered from an unknown East Coast licence area (UMA_0182). These represent isolated finds which have little or no context; in many cases the actual location of the find is unknown other than the area from which it was dredged.

Although not under discussion here, there is a potential for Early Mesolithic artefacts to be present associated with early Holocene sediment deposits. The regionally important early Holocene channel and associated Unit 7 deposits indicate the preservation of coastal deposits. The results of future management of this aggregate block will improve knowledge of the extent of the artefacts. Management and mitigation strategies are discussed in more detail below.

The Significance of the Artefact Assemblage within the Palaeo-Yare

Area 240 occupies a key position in the range of known Middle Pleistocene Palaeolithic sites in the Palaeo-Yare catchment (Table 6.2 and Fig. 6.2) and contains a relatively large Levallois component of flakes and cores. Artefacts are scarce if not absent from the onshore element of the archaeological record after MIS 8 (Wymer 1999), within the backdrop of a changing environment of warming temperatures and rising sea level, ultimately somewhat colder than the present-day.

The Area 240 assemblage contains a substantial proportion of Levallois artefacts, including 21 flakes and three cores. In comparison, there is a relative paucity of Levallois lithics within the (modern) onshore portion of the Palaeo-Yare, represented by a handful of reported Levallois flakes from sites in the upper catchment around Norwich (Wymer 1999), including the large hand axe assemblage from Keswick Mill Pit, Norwich. Future work to re-examine these assemblages and derive chronometric dates for the sites would provide valuable comparisons. The distribution of contextualised Early Middle Palaeolithic sites and hominin activity in the upper Palaeo-Yare catchment are briefly discussed by Wymer (1999) as probably reflecting some hominin presence during the early Saalian. However, the scale of the Levallois technology provides little context or predictive insight for the collection from Area 240.

This raises a significant question about the representativity of the archaeological record from terrestrial contexts: what is the potential for underrepresented, or currently unknown, periods of archaeological activity preserved within now-offshore contexts in UK waters? Certainly deposits dating to the early Devensian have been located within Area 240 and from other MALSF projects off the south and east coasts (Wessex Archaeology 2013a). Palaeogeographic models from these projects provide regional-scale models for hypothesis building and synthesis initiating the first steps in building future research priorities and project design. The continuing

Table 6.3 Faunal remains recovered from Area 240 between October 2010 and June 2012

Age	Common name	Species	Discovery dates
Early Pleistocene to Early Middle Pleistocene	Broad-fronted moose	*Alces latifrons*	October 2010 to April 2011
Early Pleistocene to Early Middle Pleistocene	Deer (giant deer)	*Megaloceros dawkinsi*	October 2010 to April 2011
Early Pleistocene to Early Middle Pleistocene	Deer	*Eucladoceros* cf. *ctenoides*	October 2010 to April 2011
Early Pleistocene to Early Middle Pleistocene	Southern mammoth	*Mammuthus meridionalis*	October 2010 to June 2012
Early Pleistocene to Early Middle Pleistocene	Horse	*Equus bressanus*	April 2011 to June 2012
Late Middle Pleistocene	Straight-tusked elephant (forest elephant)	*Palaeoloxodon antiquus*	April 2011 to June 2012
Late Pleistocene	Woolly mammoth	*Mammuthus primigenius*	October 2010 to June 2012
Late Pleistocene	Steppe bison	*Bison priscus*	October 2010 to April 2011
Late Pleistocene	Giant deer (Irish elk)	*Megaloceros giganteus*	October 2010 to June 2012
Late Pleistocene	Woolly rhinoceros	*Coelodonta antiquitatis*	October 2010 to April 2011
Late Pleistocene	Wild horse	*Equus* cf. *caballus*	October 2010 to June 2012
Late Pleistocene	Steppe bison	*Bison priscus*	April 2011 to June 2012
Early Holocene	Red deer	*Cervus elaphus*	October 2010 to April 2011
Early Holocene	Bovid	*Bos taurus*	October 2010 to April 2011

and future contribution from industry-led archaeology in support of offshore renewables, aggregates extraction, cables and pipelines is of central importance, typically incorporating higher-resolution sampling and geophysical and geotechnical assessments. Detailed mapping, palaeoenvironmental analysis and dating of geomorphological features (with clues from reported finds) are likely to continue to be the main aims of palaeolandscapes research.

Faunal Remains in Area 240 and the Southern North Sea

A large number of faunal remains have been recovered from Area 240 as part of the discovery at the wharf in 2007/2008 (around 130), during nine seabed sampling and 36 dredging monitoring operations. Of those recovered during the monitoring most were associated with mammoths and undifferentiated large herbivores and teeth including an example of a woolly mammoth (*Mammuthus primigenius*) upper molar (A. Currant pers. comm.). There was no evidence to directly associate the faunal remains with the flint artefacts (ie, no evidence of butchery marks). However, the faunal remains do indicate that the habitat was suitable for large mammals at various times throughout the Pleistocene and Holocene.

Furthermore, between October 2010 and June 2012 faunal remains (Table 6.3) were recovered at the wharf in Vlissingen by the Natural History Museum of Rotterdam in agreement with the wharf (Strijdonk *et al.* 2011; 2012); the aggregate from which this material was recovered had been dredged from Area 240. The remains contain an Early Pleistocene to early Middle Pleistocene assemblage comprising terrestrial mammal bones primarily from species of mammoth and moose. In addition, an assemblage of faunal remains dated as Late Pleistocene, which includes woolly mammoth, bison, giant deer, woolly rhinoceros and wild horse. A final group dates to the early Holocene and included red deer and bovid remains.

A further 32 faunal remains have also been reported though the Aggregate Industry *Protocol for Reporting Finds of Archaeological Interest* from the wider East Anglian aggregate dredging block. All the reported identifiable bones belonged to land mammals, including reindeer (*Rangifer tarandus*), mammoth (*Mammuthus primigenius*), giant deer (*Megaloceros giganteus*) and aurochs (*Bos primigenius*). Mammoth teeth were also reported. The majority of the finds reported in this manner are isolated finds and have little or no environmental or spatial context; in many cases the actual location of the find is limited to the extraction area from which it was dredged.

Geological Context of the Faunal Remains

The context of the recovered faunal remains is difficult to ascertain. Some of the material in the original discovery was reported to be in very good condition indicating possible *in situ* samples. However, the majority of faunal remains are abraded and rolled, and have probably been redeposited in a secondary context.

The recovery of such a large number of finds in a relatively small area is not such a surprise. Since 1874 fossil bones have been brought ashore by fishermen in the Netherlands and approximately 7500 specimens of *Mammuthus primigenius* alone have been collated by the National Museum of Natural History at Leiden (Glimmerveen *et al.* 2004). Drees (1986) documented 54 locations in the southern North Sea which have yielded at least 100 fossil bones situated to the east of the East Coast aggregate block. Although there is no geological context for these mammal faunas, based on species it is possible to group into four classes which represent terrestrial faunas which inhabited the area between Great Britain and the European Continental Shelf during different episodes of the Pleistocene and Holocene when sea levels were lowered (van Kolfschoten and Laben 1995):

I.   Early Pleistocene;
II. late Early Pleistocene/early Middle Pleistocene;
III. Late Pleistocene; and
IV. Holocene

Based on the regional geological maps there are patterns in the localities of the recovered faunal remains. Three of these four faunal association groups can be attributed to specific units within Area 240 and the wider region. Those mammals identified as late Early Pleistocene to early Middle Pleistocene (faunal association II) are likely to be associated with either Unit 2 (Yarmouth Roads Formation which is widespread throughout the southern North Sea) or Unit 3b (confined to the Palaeo-Yare Valley and not documented elsewhere). Faunal remains thought to be Late Pleistocene (Faunal association III) are likely to be associated with Unit 4, 5 or 6 in Area 240 and are associated with the Brown Bank Formation (Unit 4) on a regional scale. Early Holocene faunal remains (faunal associations IV) are likely to be associated with Unit 7 deposits off the coast of East Anglia and the Elbow Formation, remnants of which are situated in the eastern part of the Flemish Bight in the southern North Sea. However, as faunal remains tend to be light (unless heavily fossilised), it is likely that many are reworked into younger sediments including recent marine deposits that comprise the upper layers of the seabed. Early Pleistocene small-mammal remains have also been recovered from Ipswichian deposits and Holocene marine sediments, indicating reworking of the faunal remains (van Kolfschoten and Laben 1995).

The Bigger Picture

Area 240 represents transition: technologically, environmentally and palaeogeographically. By their definition, submerged prehistoric sites are more likely to derive from periods of lower-stand sea levels linked to colder transitional environments. Reconstructions of highstand or lowstand sea-level palaeogeography are of some use, but for the period of activity proposed for Area 240 (MIS 8–7), during colder, open conditions the transitional middle ground is more representative (approximately 50 m below present-day sea level, Fig. 6.2). Similar colder, open environments are recorded at most of the Early Middle Palaeolithic sites of this time in Northwest Europe (Scott and Ashton 2011). This fact alone highlights the need for an evidence-based submerged palaeolandscape to underpin all models of hominin movement between glacials and interglacials. Highstand sea-level conditions, closer to that of today represent a minimal period of time for considering Pleistocene hominins within their landscape.

Palaeogeographically, Area 240 raises a number of key questions for hypotheses on hominin colonisation of western Europe; not only for the earliest periods during the Cromerian Complex, but also for the Saalian (MIS 8/7), as well as post-glacial later Upper Palaeolithic and Mesolithic activity within the southern North Sea region. These themes are developed below.

Populating our Palaeogeography

Critically, the relative complexity of the climate and effects upon the palaeogeography of Britain and Northwest Europe during MIS 9–6 (Scott and Ashton 2011) means that contextualising the Area 240 assemblage, composed of hand axes in a fresh condition and locally eroding Levallois flakes, is a difficult task.

White (2006) has speculated on the practical methods for how Neanderthals may have recolonised Britain during the colder conditions of MIS 3 (*c.* 60 ka) after abandoning the region since MIS 7/6 (*c.* 200–140 ka). Conceptually, extending this palaeogeographic scenario of Neanderthal survival strategies within colder environments from MIS 3 back to the transition between MIS 8/7 (*c.* 250–200 ka) and to the archaeological remains recovered from Area 240 provides some parameters for developing a preliminary populated palaeogeography model.

When considering human activity – *in lieu* of informative palaeoenvironmental indicators within the sediments – defining the configuration of the landscape, at the time the artefacts were deposited, is problematic. Area 240 lies some distance to the south of more northerly limits for MIS 8 glaciation and Marine Isotope records indicate a relatively mild cold stage at this time (Scott and Ashton 2011), with sea levels as low as *c.* 100 m (Pettitt and White 2012). An open link to Northwest Europe is argued to have existed between the Thames Valley–East Anglia and the Rhine–Meuse through this entire period but with corresponding Channel routes impeded by deep water (Ashton *et al.* 2011). To what extent are differences in the type and distribution of lithic technology influenced by this palaeogeography? Further, how are different cultural groups traversing the landscape through different routes (Scott and Ashton 2011)? Assemblages of Levallois technique, and some also containing hand axes, dating to MIS 8–7 are found from north-west France towards the Rhine and Meuse catchment, and also in the Thames Valley and Suffolk. Within this context Area 240 provides a more northerly, easterly (and now-submerged) location for developing hypotheses on hominin colonisation in the Early Middle Palaeolithic. The swathe of southern North Sea and English Channel seabed corresponding to the distribution of Levallois material that may preserve further cultural material is considerable, greater than the area of Wales.

Environment and Resources

Summarising the climatic record that corresponds to Britain's early prehistoric record as a whole, Candy *et al.* (2011) have demonstrated that early human activity in Britain is associated with a broad spectrum of environments; from warmer than today, 'Mediterranean' conditions (Group 1, eg, Pakefield); comparative, temperate conditions (Group 2, eg, West Runton); cooler, temperate conditions perhaps going into or coming out of a glacial period (Group 3, eg, Happisburgh I, High Lodge and Boxgrove); and, extremely cold, periglacial conditions (Group 4, eg, Warren Hill). As a basic climatic analogue for the Area 240 site, the so-called Group 3 climate group, '*cool temperate environments, possibly reflecting the end of an interglacial or even an interstadial*' (Candy *et al.* 2011, 19), is the most similar.

Generalised glacial-interglacial floral development would suggest a basic framework on which we might pin our interpretations. During a cooling trend leading into a glacial period, coniferous forests would open into grassland, which in turn would develop into a dry open woodland as conditions warmed leading into an interglacial (Leroy *et al.* 2011). The limited floral and microfaunal records (with reworked charcoal and wood) preserved within vibrocores from Unit 3b in the Area (VC29, Area 254; Limpenny *et al.* 2011) indicates a cold estuarine environment with a relative sea level around 25–30 m below OD.

The environmental pressures driving a need for clothing and artificial shelter discussed by White (2006) are perhaps not as restrictive during MIS 8/7 (compared to MIS 3) due to a more elevated climatic amelioration towards MIS 7e (*c.* 240 ka) (Scott and Ashton 2011; Ashton *et al.* 2011), still within a lower sea-level palaeogeography. Perhaps winter temperatures would have proved difficult. Area 240's latitude is notably further north than other Northwest European sites at this time; it would have provided a less variable, maritime climate, though without the protective cave environment of Pontnewydd, Clwyd, which is located further north. There is also scope for speculating on possible seasonality of occupation or pursuit of resources when conditions allowed.

The geophysical, geotechnical and archaeological evidence from Area 240 as a whole indicates an ecotone: a coastal, estuarine location near the banks of river on the margins of grassland, probably cool steppe landscape with some trees during late MIS 8/early MIS 7. Nearby, the Anglian till cliff would have framed the coastal landscape in a similar topography to the present-day coastline. Relative sea-level rise during this ameliorating phase and marine transgression would have acted to modify the relative configuration of coastal environs and resources over time.

It is difficult to discern whether this landscape was populated by large mammals favoured by Neanderthal hunters (Scott and Ashton 2011; Pettitt and White 2012) based on the recovered faunal material from the site. This is mainly due to problems with direct dating of organic material beyond the limits of radiocarbon dating and relatively poorly constrained sedimentary contexts for dredged materials (although well over a hundred finds of faunal material have been recovered to date). The recovery of identifiable mammoth remains lends some support to mammoths being a substantial element in the environment and potentially hominin subsistence (Pettitt and White 2012) in the vicinity of Area 240.

Movement and Colonisation

It is acknowledged that archaeologists seek to understand any given early prehistoric site beyond the mere presence of lithics and the technical details of an assemblage. Indeed, implicit to site investigation is the goal (sometimes unachievable) to investigate the nature, environment and economies of the prehistoric peoples we seek to study, and understand. A key result that eludes us from Area 240 is the volume and extent of the archaeological resource and a full sample of the lithic assemblage (including debitage which is preferentially screened out during wharf processing), tightly contextualised by contemporary faunal and floral remains. The lithic assemblage offers few clues to the duration, extent and development of human activity. Does Area 240 reflect a place attractive to humans over a long period? Or, is a particular site with one or more short phases of activity represented? Is this linked to seasonal or annual modes of subsistence and/or raw material gathering and lithic production?

Scott and Ashton (2011) suggest that the later warming limb of MIS 8 could have supported habitats suitable for human colonisation; lasting approximately 10,000 years and dating roughly to 250–240,000 years ago. This is comparable to the available dating evidence from the site Area 240 and also recent speculative models of human colonisation into Britain during the Saalian (Ashton *et al.* 2011: fig. 4.2). These focused parameters allow for future research questions at Area 240: and thematic questions that should be applied to submerged prehistoric sites discovered in the future. Whilst we may view a colonisation window of 10,000 years (or survival) between 250–240 ka as brief on geological timescales, on a human scale, this 10,000 years is the equivalent of several thousand generations – arguably enough substantial time for expansion the migration of Neanderthals from Northwest Europe into southern England during late MIS 8/early MIS 7.

Palaeogeographical reconstructions place Area 240 close to the Saalian coast of East Anglia, populated by Neanderthals. Happisburgh and Pakefield located near to Cromerian Complex coasts (which happen to correspond to the modern coastline) were populated by more ancient hominins across a gulf of warmer (than now) and colder (than now) climatic conditions. Within the context of other Early Middle Palaeolithic sites in Northwest Europe, coasts do not commonly appear close to site locations (Scott and Ashton 2011; Pettitt and White 2012). However, coastal plains are key environments for hunting large mammals and general subsistence, much of this biotope is now under water, with the terrestrial archaeological record located on the landward side of the relict plain or in the upper catchment such as the case for the Palaeo-Yare but similar factors underpin the palaeolandscape context of Lower Palaeolithic sites and environs such as Boxgrove (West Sussex), and Valdoe Quarry, West Sussex, respectively, on raised beaches overlooking the coastal plain (James *et al.* 2010; James *et al.* 2011; Pope 2010; Cohen *et al.* 2012).

Within the Palaeo-Yare catchment the apparent restriction of Middle Palaeolithic hominin activity with a substantial Levallois component to the lower reaches of the river, might reflect hominin subsistence concentrated in coastal fluvial plains where venturing further inland did not significantly occur or was not sufficient to leave a substantial archaeological record. However, factors, other than taphonomy and issues of sampling, may be prevalent.

Erlandson (2001) has critiqued the use of marine resources for earlier prehistory arguing for an underrepresentation in models of hominin subsistence which are certainly important for Mesolithic hunter-gatherers (*cf.* Bailey and Spikins 2010). Recent work has placed coasts and an 'Atlantic climate' firmly at the heart of interpreting the distribution of the earliest human occupation of Britain (Cohen *et al.* 2012). In southern Britain, the archaeological record for establishing the importance of more coastal areas during the last one million years of prehistory is likely to be under water. Therefore, any debate or investigation into research topics of this nature will now be required to push beyond mere *potential* and there will be no acceptable reason to overlook or discard what is now seabed. Rather, any critical, evidence-based approach will incorporate environmental and (geo)archaeological evidence into broader discussions of British, and indeed world prehistory (*cf.* Flemming 2004; Benjamin *et al.* 2011).

Method Evaluation

The evidence-based approach of the project has successfully investigated Area 240 for the presence of further artefacts and has established the prehistoric character of the wider region. A range of methodologies was applied to gauge their effectiveness in identifying and assessing sites of this type. These methods are discussed below in conjunction with a discussion on future work in Area 240 (and beyond).

Reconstruction of the Palaeogeography

The palaeogeography of Area 240 has been characterised successfully using a combination of geophysics and geotechnical datasets, vibrocore acquisition and geoarchaeological logging and palaeoenvironmental assessment, analysis and dating (radiocarbon and OSL). The combined approach resulted in a comprehensive interpretation of the stratochronology of Area 240. This project has afforded the time and development such that a much more detailed interpretation of Area 240 has been accomplished in comparison to the interpretation typically conducted during the course of an Environmental Impact Assessment (EIA) for an aggregate assessment.

Further work, however, could serve to improve the palaeogeographic reconstruction and understanding of the Area. Only four vibrocores were palaeo-environmentally analysed as part of the project. Further analysis of the stored cores would give further confidence in the interpretation and is possible that more palaeoenvironmental remains may be garnered from Unit 3b furthering information on the Saalian environment. Also, the age and environment of the infilled depressions (Unit 5) is still outstanding at this time. Any vibrocores acquired in the future should be archaeologically assessed and if appropriate sediment samples should be palaeoenvironmentally analysed and dated. This would complete the stratigraphy within Area 240 and the dating would further clarify the hypotheses regarding the likely scenarios concerning the geological context described earlier.

The stratigraphy observed in the geophysical and sediment data was stratigraphically constrained by OSL dating. Without the dating it would have been nearly impossible to attribute MIS stages to the sediment units. However, a major limitation, especially with older sedimentary bodies, has been the integrity of OSL samples from vibrocores. Due to the potentially complex taphonomy of offshore sediments, on the rare occasion that OSL dating of sediments has been applied, the resulting dates have been of variable applicability. Where radiocarbon dating cannot be used either due to age constraints or lack of organic components, OSL dating must typically be applied. To improve the quality of dates improved sampling strategies must be adopted to identify unmixed and well-bleached sediments that will provide greater understanding of sediment taphonomy, increased cost-effectiveness of dating campaigns, ultimately greater confidence in age models and, therefore, interpretations. It is especially important to maximise the financial investment in establishing a chronology during development-led projects and tools exist for these purposes (Sanderson and Murphy 2010).

The OSL dating results from Area 240 indicate the clear potential for further detailed chronological assessment by luminescence techniques to provide improved precision and taphonomic understanding. The application of single-grain techniques and OSL profiling would help to elucidate the distribution of sensitive, bleached sediments most suitable for OSL dating, and unbleached or partially-bleached, reworked sediments which degrade accuracy and precision (Sanderson and Murphy 2010). Additional statistical analysis (eg, Cunningham and Wallinga 2012) may also be beneficial. With the substantial quantity of stored, dark-sampled vibrocore sections (archived by Wessex Archaeology), significant and cost-effective geochronological work could be easily undertaken and integrated with palaeoenvironmental analyses.

By expanding the frequency and scope of palaeoenvironmental sampling and especially dating deposits stratigraphically (above and below deposits of archaeological interest), dating control can be substantially improved. Very few assessments from offshore contexts have proceeded to detailed palaeoenvironmental analysis and dating; most stop at a general characterisation based upon published terrestrial sources or regional geological references. This largely precludes any step devoted to detailed correlation with existing knowledge (except at local scales) and limits syntheses to regional overviews.

On a regional scale, further palaeoenvironmental analysis and dating of cores acquired during the East Coast REC could further inform the setting of the Palaeo-Yare within the wider region. It is clear from the dating evidence that the sediment preservation in this part of the southern North Sea is more complex than previously thought. Also, palaeoenvironmental analysis and dating of vibrocores acquired from the remaining licence areas within the East Coast aggregate block are recommended in the future to constrain the nature, age and extent of the Palaeo-Yare floodplain deposits.

Dredging over the past three decades means that the state of preservation of the sediment deposits through natural process alone is unknown. As such it is unclear what quantity of artefacts, if any, have been dredged previously and have been processed as aggregate. Similarly, dredging has continued throughout much of Area 240 since the main dataset was acquired in 2005 (excluding the AEZ since 2009). Interpretation of any geophysical data acquired in the future at Area 240, or in the surrounding aggregate block, may help to indicate the potential remnants of Unit 3b and likely locations for the presence of artefactual material.

Sampling Artefacts

The methods utilised in Area 240 for sampling of artefacts were specific to the conditions of the site and applied established techniques used within the EIA process. The advantage of applying existing techniques for archaeological purposes was that deployment, recovery and processing are well established and integration between archaeological use and existing use was found to be straightforward. As discussed in Chapter 5 the success of these techniques was variable, though given the recovery of flint artefacts and faunal remains, they proved to be successful. Although artefacts were recovered by the Hamon grab there were two main issues with this methodology: 1) the volume of recovered sediment was too little and; 2) it was difficult to target the buried sediment unit within which the artefacts were preserved.

The use of clamshell grab (280 litres) was an improvement on the Hamon grab (10 litres) as used in previous archaeological trials; the increase in the size of the equipment not only resulted in sampling a larger volume of sediment but also penetrated the seabed sediments to approximately 0.5 m sub-seabed compared to 0.1 m using the Hamon grab. Although a clamshell grab penetrates the seabed by 0.5 m the artefacts that were recovered using this method were very abraded and based on the sediment composition were thought to be from the uppermost marine sediment unit (Unit 8). Although accurate positional control was achieved during the survey, greater vertical control over the sample would have been beneficial in order to target specific sediment units.

In order to successfully sample buried targets, existing equipment may need to be specifically redesigned. For example, the use of a large suction pump regularly used by divers could potentially be adapted to provide positioning (vertical and horizontal) control. With adequate control it may be possible to remove a specific sediment sample from the seabed, and sieve the contents for artefacts before acquiring the next sample. An approach focusing on this removal of small samples has recently been used during development activities in the port of Rotterdam (Weerts *et al.* 2012). Artefacts, including fragments of burnt animal bone and fragments of flint and flint flakes dating to the Early and Middle Mesolithic (9–11 ka) have been located in 20 m water depth. From pontoons, grabs systematically scrape slices of sediments in 0.2 m layers from the seabed. These samples are then sieved for traces of artefacts and environmental remains. Although this specific technique would not applicable for the Area 240 site (given the exposed, open-sea conditions), this approach ensures greater vertical control over the sampling. This is a creative example of tailoring fieldwork to assess an individual submerged site.

Through the targeted dredging and monitoring of the load, we furthered our experience, tested different sampling techniques and ensured an increase of sediment volume was assessed. A total of 24 flint artefacts were recovered using this method. There are very little comparable data to demonstrate how the rates of artefact recovery using this method may indicate potential densities of material from the gravel. However, observations made during a terrestrial archaeological watching brief, undertaken at a gravel pit in the River Test Valley at Dunbridge (the most prolific Palaeolithic location in Hampshire), may be comparable (Harding *et al.* 2012). A field notebook for this site records that on 28 April 2006 one broken and heavily rolled hand axe was recovered from the (40 mm) reject heap, estimated as 3000 tons, (the equivalent of 60% of a dredger cargo). In comparison, the 5000 tons of Trip 1 during the 2011 trial dredging yielded 11 flint artefacts. Drawing such comparisons is of some, albeit limited, value due to methodological variations however it remains relevant and, therefore, of interest. The use of a dredger as a large sampler is an option for similar sites to Area 240 which are located in aggregate extraction areas. However, this technique is not likely to be practical for assessing sites in other contexts. Furthermore, although not applicable for Area 240, more shallow or sheltered coastal locations diving could be a suitable methodology, although locating buried artefacts would remain problematic.

Assessing archaeological sites in the offshore environment is challenging; archaeologists may have to look to other methods such as the use of remotely operated vehicles (ROVs). The technical issues that arise from underwater research are best confronted through interdisciplinary collaboration: archaeologists working with other marine scientists to investigate a wide range of techniques or adapting established methods and equipment previously untested for their archaeological application.

Summary

It is clear that there is no single methodology, or sequence of methodologies, that can be recommended for use at all underwater archaeological sites. Each site presents its own real-world challenges and therefore, must be treated on its own merits with sampling strategies and methods tailored to suit its particular conditions. However, a combination of methods including geophysics, geotechnical coring (with associated analysis and dating), diving, and sediment sampling should all be considered when planning such work. The development of technologies such as the use of targeted dredging and ROV's should also be considered.

Due to the methods and complexity in investigating submerged prehistoric sites like Area 240, interpretations and reconstructions are necessarily restricted to larger spatial and temporal scales. The landscape and overall distribution and context of artefacts can be reconstructed to varying degrees, however critical archaeological questions regarding human migration and response to climatic and environmental change, must be framed by appropriate scale and precision. Relatively young, non-glaciated, shallowly submerged, non-industrial sites such as Bouldnor Cliff (Momber *et al.* 2011) and the post-glacial sites from the south-west Baltic (*cf.* Lübke 2011; Fischer 2011) can be examined in detail by divers allowing human-scale reconstructions. At Area 240, by maintaining an evidence-based approach to content and interpretation, we must be satisfied with a broader picture of landscape changes, environmental evolution and a general analysis of human activity based on our knowledge of the lithics introduced into the archaeological record and their associated industries.

In the discussions above, appropriate techniques for sampling known archaeological sites are examined. Prospecting for, and finding, unknown, archaeological sites under water is much more challenging. This is discussed further below.

Management and Mitigation

Since the recovery of the initial artefacts from Area 240 a series of management strategies have been employed. The implementation of the AEZ by Hanson Aggregate Marine Limited protected the area from further dredging and allowed the undertaking of a detailed assessment of the site. Subsequently, monitoring of Area 240 dredge loads took place with the aim of assessing potential mitigation strategies in a region of known archaeology, with regards to future long-term aggregate licensing applications.

Based on the findings, both of this work and the subsequent 2011 monitoring programme, it was acknowledged by the aggregate industry that the relationship between the apparently *in situ* archaeological material and the geological context could not effectively be carried out on a licence-by-licence area basis and a regional framework was required to manage the aggregate block with a common management strategy. Following an extensive assessment of the Palaeo-Yare catchment, focusing on the extent of the floodplain and Unit 3b in particular, a series of hypotheses were proposed aimed at elucidating both archaeological and management questions; such as, does the distribution of archaeological material reflect human choice or taphonomic processes? Does intensive aggregates dredging preclude the finding of future prehistoric materials in the future? It is anticipated that the information resulting from testing these hypotheses by dredge monitoring will enhance the knowledge of the presence of Palaeolithic material in the wider area and will inform the future management of the aggregate block as a whole. This approach is unique and has been specifically developed for an area in which a known archaeological resource has been studied. At present, other aggregate areas are only discussed in terms of their potential for prehistoric archaeology. However, if further archaeological assemblages are found in the future in other aggregate areas, the archaeological and management strategies employed at Area 240 may be used as a starting point. However, it is acknowledged that all such sites would vary and as such assessment strategies would need to be tailored to each site; no one strategy will be appropriate for all scenarios.

Further Afield

For archaeological assessments in support of the aggregates industry, the knowledge base for understanding the development and context of palaeolandscapes and their significance for prehistoric archaeology, are generally well developed for all regional licence areas. Through the examination of geophysical and geotechnical datasets specific palaeogeographical and archaeological assessments can be made, avoiding the need for generalised statements of potential and facilitating the effective management of both the aggregates resource and the archaeological resource, if present. However, palaeoenvironmental analysis and dating are not generally conducted as part of EIA and, in certain geological areas this added information is required to fully elucidate the palaeogeographical interpretation of an area. This was clearly the case at Area 240.

The successful implementation of the Marine Aggregate Industry *Protocol for Reporting Finds of Archaeological Interest* and the Offshore Renewables Protocol for Archaeological Discoveries (ORPAD) has increased the recovery and reporting of archaeological material. As more and more material is reported distribution patterns can be analysed and integrated into palaeolandscape interpretations. The implementation of a similar scheme for the fishing industry could serve to expand this knowledge.

Very little baseline information on submerged prehistoric palaeolandscapes within British waters existed, even as recently as the end of the 20th century. This knowledge base has developed substantially during the past decade and has expanded to the point where national syntheses are now required (Wessex Archaeology 2013a). The benefit to industry and sustainable development is significant. For development in areas where substantial research has already been completed (eg, southern North Sea and East English Channel), baseline conditions are well known and informed EIA can be undertaken effectively. In relatively un-investigated areas, such as Round 3 renewable energy zones, the methods and experience developed during the last decade, are invaluable and will ensure that prehistoric archaeology is considered and examined to the highest standard.

Within UK waters there is characteristic palaeolandscapes potential within and between regions (Wessex Archaeology 2013a). At a regional scale, relict landforms and sediments contemporary with the Mesolithic have been identified around most coasts within British waters with some, substantial offshore areas ground-truthed by development-led assessments and research projects (eg, Gaffney *et al.* 2007; Fitch *et al.* 2011).

Prior to the Holocene, regional palaeolandscape configurations vary for Palaeolithic archaeological periods. For example, the Irish Sea appears especially characterised by post-glacial later Upper Palaeolithic palaeolandscape features. During the last 500 ka the important terrestrial Lower Palaeolithic archaeological record of the south coast has been contextualised by regional palaeolandscape reconstructions (James *et al.* 2010; 2011). Broadening the investigation of post-glacial, Upper Palaeolithic submerged palaeolandscapes, and the sparse terrestrial record of various hunter-gatherer cultures sporadically recorded from southern Scotland to southern England, would be archaeologically important across 'Doggerland'.

Such models provide regional-scale resources for developing research questions and testing offshore potential for earlier hominin activity. Does potential for archaeological material, similarly aged to Happisburgh and Pakefield, exist in the south coast region, especially towards the East English Channel? A number of questions can be formalised for offshore locations where evidence is scarce, equivocal or not preserved within terrestrial contexts – a fundamental question which underlies existing prehistoric synthesis is, how biased is the archaeological record preserved in terrestrial contexts? Regardless of which glacial or interglacial period is under scrutiny, hominin routes through the landscape, the strategies used to procure food, materials and shelter, are key archaeological components to questions that should be factored into future research and informed by palaeogeographical models (to include now-submerged environments).

Substantial evidence-based palaeolandscape reconstructions now exist especially for Lower, Middle and Upper Palaeolithic and Mesolithic periods in the Channel and North Sea regions enabling further hypotheses for the colonisation of Britain by *Homo antecessor*, *Homo heidelbergensis*, *Homo neanderthalensis* and *Homo sapiens* across several broad routes: a North Sea coastal route, an Atlantic coastal route, the Weald-Artois ridge itself or a combination (Cohen *et al.* 2012; Scott and Ashton 2011; Ashton *et al.* 2011). For example, if the artefacts at Happisburgh are made by *H. antecessor* how did the diaspora develop through Europe to reach this position at 'the end of the world'? Was this an Atlantic route from Northern Spain or through Northwest Europe and across the North Sea coasts (Cohen *et al.* 2012)? The emerging picture of submerged palaeolandscapes (Wessex Archaeology 2013a) is that despite concerns about preservation bias (Cohen *et al.* 2012), sediments of archaeological interest relevant to the Lower Palaeolithic, and every major archaeological period until at least the Iron Age, have been located within offshore, nearshore and intertidal areas of British waters. Many of these geomorphic features – particularly palaeochannels, river terraces, estuaries, marshes and peats – are related to flint-rich fluvial systems and temperate climates. The vast majority of this submerged palaeolandscape corpus has been produced by the development-led archaeology sector and MALSF projects (Wessex Archaeology 2013a).

Critically, notable gaps within the British prehistoric record provide opportunities for investigating themes for which only submerged contexts may preserve the physical evidence. Submerged palaeolandscapes may preserve critical context for understanding these periods of absence or recolonisation when reconciling the existing terrestrial evidence has proven to be problematic, especially during the Ipswichian interglacial (Lewis *et al.* 2011), but also for other periods (Pettitt and White 2012). Are palaeolandscapes dating to the warming or cooling limb of MIS 5e preserved in offshore contexts? Palaeolandscape reconstructions for MIS 5a–d (*c.* 120 ka) in the East English Channel (James *et al.* 2010; 2011) also hint at avenues for developing future research into human absence from Britain during the Ipswichian/early Devensian where evidence from the French side shows sites within the river valleys near the current coast (James *et al.* 2011). MIS 3 dates from Area 240 suggest this important recolonisation period for Neanderthals around 60–40 ka could be focused on now-submerged contexts, building on speculative models (White 2006). Similarly, the terrestrial evidence for Neanderthal and modern human activity during this time is diagnostically complex and at the limits of radiocarbon dating techniques (Jacobi and Higham 2011). Major research questions from offshore contexts can be readily assembled from the current knowledge base; perhaps clarifying the British evidence for habitation during MIS 6–4 in addition to testing models on palaeogeographic and technological barriers to access (Ashton *et al.* 2011).

Although there is now a much greater understanding of submerged landscapes on a regional scale there is still a knowledge gap regarding the quantity and location of archaeological sites within these palaeolandscapes. The scale of sampling, analysis and interpretation required for development-led, and most MALSF projects, has focused on larger areas or regional scales. However, despite the increased volume of data, the prediction of sites within such expansive submerged landscapes is difficult; this relatively coarse resolution alone is unlikely to successfully locate individual archaeological sites. Although this has been successfully accomplished in the relatively straightforward palaeogeographical situation in the Baltic for locating post-glacial sites in nearshore environments (Fischer 1995); this is not the case for older, more complex Pleistocene landscapes. Resolving the chronology of identified features and moving beyond accounts of landscape remains a distinct challenge (Sturt and Standen 2013). The submerged prehistory research community acknowledge that consideration needs to be given to direct sampling of offshore deposits with a view to identification of archaeological material. However, it is not currently feasible to systematically search submerged landscapes for pre-Holocene sites in British waters due to the spatial scale under consideration, the expense of underwater investigations and the variability of the underwater environment. The numerous considerations determining the archaeological potential negate the possibility for a simplified approach to investigation that does not consider detailed geophysical and geotechnical studies (Benjamin 2010), particularly in the offshore environments of the North Sea. Although targeted sampling for archaeological material should be part of current and future research agendas, given the coasts and available resources fortuitous discovery of archaeological sites will continue and will require appropriate response. In offshore areas (as opposed to coastal, nearshore environs), this will likely be undertaken through working with industry. The work carried out at Area 240 has highlighted what can be achieved through collaboration between archaeologists, heritage managers (regulators) and industry.

It is now entirely feasible to expand our existing understanding of submerged prehistory and palaeolandscapes through new research, and to prospect into new areas. This will enable integration with similar research from mainland Europe to produce a fully *source-to-sea* approach to prehistoric archaeology and palaeogeography; where modern coastlines are no boundary to theories and concepts. This should be a major priority for the next generation of research.

Key Conclusions

The considerable work from Area 240 is important for a range of archaeological, palaeogeographical and management reasons:

- The Early Middle Palaeolithic assemblage from Area 240 has survived multiple phases of glaciation and marine transgression (and regression) negating generalised assumptions that submerged prehistory in UK waters has been entirely destroyed by these factors;
- The multi-disciplinary, evidence-based approach of this project has been used to successfully investigate Area 240 for the presence of further artefacts and to establish the prehistoric character of the area;
- A range of methodologies was applied to gauge their effectiveness in identifying and assessing submerged and buried sites. The combination of geophysical, geotechnical, seabed sampling and a variety of palaeo-environmental analysis and dating techniques were used. It is critical that assessment strategies take into account the local geological and environmental conditions; one strategy does not fit all sites;
- Archaeologically, as an Early Middle Palaeolithic assemblage the rarity and associated scientific value of the site is very high (Pettitt *et al.* 2008); as the only offshore, submerged Early Middle Palaeolithic assemblage it is important within the context of world prehistory.

The results have shown that submerged landscapes can preserve *in situ* Palaeolithic artefacts. The investigations confirm that the artefacts are not a 'chance' find, but indicate clear relationships to submerged and buried landscapes that, although complex, can be examined in detail using a variety of existing and new fieldwork and analytical methods. The process was successful, yielding positive results when testing methods designed to encounter submerged landscapes and *in situ* prehistoric material. Through close collaboration between archaeologists, regulators and industry it has been possible to go beyond an assessment of potential submerged prehistory and identification of buried geomorphological features, and investigate the archaeology.

Bibliography

Allen, R.G. and Sturdy, R.G., 1980 The environmental background in D.G., Buckley (ed.), *Archaeology in Essex to AD 1500*, London, Counc Brit Archaeol Res Rep 34, 1–7

Alluvial Mining Ltd, 1999 *Aggregate Prospecting Survey. Areas 240 and 240B*, unpubl rep 96524–1

Andel, T.H. van, 1989 Late Quaternary sea-level change and archaeology, *Antiquity* 63, 733–45

Andel, T.H., van, 1990 Addendum to "Late Quaternary sea-level changes and archaeology", *Antiquity* 64, 151–2

Andersen, S.H., 2011 Ertebolle canoes and paddles from the submerged habitation site of Tybrind Vig, Denmark, in J. Benjamin, C. Bonsall, C. Pickard, and A. Fischer (eds), *Submerged Prehistory*, Oxford and Oakville, Oxbow Books, 1–14

Andrews Survey, 2000a *Area 240 Vibrocore Survey. March 2000*, unpub rep

Andrews Survey, 2000b *Area 240 Vibrocore Survey. July 2000*, unpubl rep

Andrews Survey, 2005 *Areas 240, 242 and 328. Vibrocore Survey*, unpubl rep

Antoine, P., Coutard, J.-P., Gibbard, P., Hallegouet, B., Lautridou, J.-P., and Ozouf, J.-C., 2003 The Pleistocene rivers of the English Channel region, *J Quat Sci* 18 (3–4), 227–243

Arthurton, R.S., Booth, S.J., Morigi, A.N., Abbott, M.A.W., and Wood, C.J., 1994 *Geology of the Country around Great Yarmouth*, London, HMSO

Ashton, N., 1992 The High Lodge flint industries London, in Ashton 1992, 124–163

Ashton, N., 2002 Absence of humans in Britain during the last interglacial (Oxygen Isotope Stage 5e), in A. Tuffreau and W. Roebroeks (eds), *Le Dernier Interglaciaire et les Occupations Humaines du Palaeolithique Moyen*, Lille, Publications du CERP, 93–103

Ashton, N., Cook, J., Lewis, S.G. and Rose, J., 1992 *High Lodge: Excavations* by G. de G. Sieveking, 1962–68 and J. Cook 1988, London, British Museum Press

Ashton, N. and Hosfield, R., 2009 Mapping the human record in the British early Palaeolithic: evidence from the Solent River system, *J Quat Sci* 25 (5), 737–753

Ashton, N. and Lewis, S., 2002 Deserted Britain: declining populations in the British Late Middle Pleistocene, *Antiquity* 76, 388–396

Ashton, N., Lewis, S.G., De Groote, I., Duffy, S.M., Bates, M., Bates, R., Hoare, P., Lewis, M., Parfitt, S. A., Peglar, S., Williams, C., and Stringer, C., 2014 Hominin footprints from Early Pleistocene deposits at Happisburgh, UK, *PLOS ONE* 9 (2), e88329

Ashton, N., Lewis, S., Parfitt, S., Candy, I., Keen, D., Kemp, R., Penkman, K., Thomas, G., Whittaker, J. and White, M., 2005 Excavations at the Lower Palaeolithic site at Elveden, Suffolk, UK, *Proc Prehist Soc* 71, 1–61

Ashton, N. and Lewis, S.G., 2012 The environmental contexts of early human occupation of northwest Europe: the British Lower Palaeolithic record, *Quat Int* 271, 50–64

Ashton, N., Lewis, S.G. and Hosfield, R., 2011 Mapping the human record: population change in Britain during the Early Palaeolithic, in Ashton *et al.* 2011, 39–51

Ashton, N., Lewis, S.G., Parfitt, S.A., Penkman, K.E.H. and Coope, G.R., 2008 New evidence for complex climate change in MIS 11 from Hoxne, Suffolk, UK, *Quat Sci Rev* 27 (7–8), 652–668

Ashton, N., Lewis, S., Parfitt, S. and White, M.J., 2006 Riparian landscapes and human habitat preferences during the Hoxnian (MIS 11) Interglacial, *J Quat Sci* 21 (5), 497–505

Ashton, N., Lewis, S.G. and Stringer, C., (eds), 2011 *The Ancient Human Occupation of Britain*, Developments in Quaternary Science 14, Amsterdam, Netherlands, Elsevier B.V, 39–51

Ashton, N. and McNabb, J., 1994 Bifaces in perspective, in N. Ashton and A. David (eds), *Stories in Stone*, Lithic Studies Society Occ Pap 4, 182–191

Athersuch, J., Horne, D.J. and Whittaker, J.E., 1989 *Marine and Brackish Water Ostracods*, Synopses of the British Fauna (New Series), Linnean Society of London and the Estuarine and Brackish-water Sciences Association, Synopses of the British Fauna No. 43

Austin, L., 1997 Palaeolithic and Mesolithic, in J. Glazebrook (ed.), Research and Archaeology: *A Framework for the Eastern Counties, 1. Resource Assessment, East Anglian Archaeology*, Occasional Paper, 3, Norwich, The Scole Archaeological Committee for East Anglia, 5–11

Baggaley, P.A. and Arnott, S.H.L., 2013 Geoarchaeology of the Channel/Manche:

combining regional and localised investigations, in Daire *et al.* 2013, 573–582

Bailey, G., 2004 The wider significance of submerged archaeological sites and their relevance to world prehistory, in Flemming 2004, 3–10

Bailey, G. and Flemming, N., 2008 Archaeology of the continental shelf: marine resources, submerged landscapes and underwater archaeology, *Quat Sci Rev* 27 (23–24), 2153–2165

Bailey, G. and Sakellariou, D., 2012 SPLASHCOS: submerged prehistoric archaeology and landscapes of the Continental Shelf, *Antiquity* 86, 334

Bailey, G. and Spikins, P. (eds), 2010 *Mesolithic Europe*, 2nd edn, Cambridge, Cambridge Univ Press

Barrett, J.H. and Yonge, C.M., 1958 *Collins Pocket Guide to the Sea Shore*, London, Collins

Barton, N., 2005 *Ice Age Britain*, London, Batsford and English Heritage

Battarbee, R.W., 1986. Diatom analysis, in B.E. Berglund (ed.), *Handbook of Holocene Palaeoecology and Palaeohydrology*, 527–570, Chichester, John Wiley

Battarbee, R.W., Jones, V.J., Flower, R.J., Cameron, N.G., Bennion, H.B., Carvalho, L., and Juggins, S., 2001 Diatoms, in J.P. Smol and H.J. Birks (eds), *Tracking Environmental Change Using Lake Sediments Volume 3: Terrestrial, Algal, and Siliceous Indicators*, Dordrecht, Kluwer Academic Publishers, 155–202

Beets, D.J., Meijer, T., Beets, C.J., Cleveringa, P., Laban, C. and Van der Spek, A. J.F., 2005 Evidence for a Middle Pleistocene glaciation of MIS 8 age in the southern North Sea, *Quat Int* 133–134, 7–19

Bellamy, A.G., 1998 *The UK marine sand and gravel dredging industry: an application of Quaternary geology*, Geological Society, London, Engineering Geology Special Publications 13 (1), 33–46

Benjamin, J., 2010 Submerged prehistoric landscapes and underwater site discovery: reevaluating the "Danish Model" for international practice, *J Island and Coastal Archaeol* 5, 253–270

Benjamin, J., Bonsall, C., Pickard, C. and Fischer, A., 2011 *Submerged Prehistory*, Oxford, Oxbow Books

Bennett, K.D., 1994 *Annotated catalogue of pollen and pteridophyte spore types of the British Isles*, unpubl rep for University of Cambridge

Bennett, K.D., Whittington, G. and Edwards, K.J., 1994 Recent plant nomenclatural changes and pollen morphology in the British Isles, *Quat Newslett* 73, 1–6

Bicket, A., 2011 *Submerged Prehistory: Research in Context*, MALSF Science Monogr. 5

Bicket, A., Firth, A., Tizzard, L. and Benjamin, J., 2014 Heritage Management and Submerged Prehistory in the United Kingdom, in J. Flatman, J.G. Evans, and N. Flemming (eds), *Prehistoric Archaeology of the Continental Shelf: A Global Review*, New York, Springer, 213–32

Boëda, E., 1984 Méthode d'étude d'un nucleus Levallois à l'éclat preferential, *Cahiers géogr. physique* 5, 95–134

Boëda, E., 1986 Le débitage de Biache-Saint-Vaast (Pas de Calais): première étude technologique, *Bulletin Association Française pour l'Etude du Quaternaire suppl.*, 26, 209–218

Boëda, E., 1988 Le concept Levallois et évaluation de son champ d'application, in M. Otte (ed.), *L'homme de Néandertal, 4: La technique*, Liege, Etudes et recherches archéologiques de l'Université de Liège 31, 13–26

Boëda, E., 1994 *Le Concept Levallois: Variabilité des Méthodes* (Monographie De CRA. 9), Paris

BMAPA and English Heritage, 2003 *Marine Aggregate Dredging and the Historic Environment: Assessing, evaluating, mitigating and monitoring the archaeological effects of marine aggregate dredging*, London, English Heritage

BMAPA and English Heritage, 2005 *Marine Aggregate Dredging and the Historic Environment*, London, English Heritage

Boismier, W., Gamble, C.S., and Coward, F., (eds), 2012 *Neanderthals Among Mammoths: Excavations at Lynford Quarry, Norfolk*, Swindon, English Heritage

Boismier, W., Schreve, D.C., White, M.J., Robertson, D.A., Stuart, A.J., Etienne, S., Andrews, J., Coope, G.R., Field, M., Green, F., Keen, D.H., Lewis, S.G., French, C.A., Rhodes, E., Schwenninger, J.-L., Tovey, K. and O'Connor, S., 2003 A Middle Palaeolithic site at Lynford Quarry, Munford, Norfolk : interim statement, *Proc Prehist Soc* 69, 314–324

Boomer, I. and Godwin, M.E., 1993 Palaeoenvironmental reconstruction in the Breydon Formation, Holocene of East Anglia, *J Micropalaeontology* 12, 35–46

Bordes, F., 1961 *Typologie du Paleolithique ancien et moyen*, Paris, Centre National de la Recherche Scientifique

Bordes, F., 1972 *A Tale of Two Caves*, London, Harper and Row

Boyd, S.E., 2002 *Guidelines for the Conduct of Benthic Studies at Aggregate Dredging Sites*, London and Lowestoft, UK Department of Transport, Local Government and the Regions and Centre for Environment, Fisheries and Acquaculture Science

Briant, R.M. and Bateman, M.D., 2009 Luminescence dating indicates radiocarbon age underestimation in late Pleistocene fluvial deposits from eastern England, *J Quat Sci* 24, 916–927

Bridgland, D., 1994 The Pleistocene of the Thames, in D. Bridgland (ed.), *Quaternary of the Thames*, Geological Conservation Review Series No. 7, London, Chapman and Hall

Bridgland, D. and Westaway, R., 2008 Climatically controlled river terrace staircases: a worldwide Quaternary phenomenon, *Geomorphology* 98 (3/4), 285–315

Bridgland, D.R., 2002 Fluvial deposition on periodically emergent shelves in the Quaternary: example records from the shelf around Britain, *Quat Int* 92, 25–34

Bridgland, D.R., Harding, P., Allen, P., Candy, I., Cherry, C., Horne, D.J., Keen, D.H., Penkman, K.E.H., Preece, R.C., Rhodes, E.J., Scaife, R., Schreve, D.C., Schwenninger, J.-L., Slipper, I., Ward, G.R., White, M.J., White, T.S. and Whittaker, J.E., 2012 An enhanced record of MIS 9 environments, geochronology and geoarchaeology: data from construction of the High Speed 1 (London–Channel Tunnel) rail-link and other recent investigations at Purfleet, Essex, UK, *Proc Geol Ass* 124 (3), 417–476

British Geological Survey (BGS), 1991 *East Anglia Sheet 52°N – 00°. 1:250000 Series Quaternary Geology*, Natural Environmental Research Council

Brock, F., Higham, T., Ditchfield, P. and Bronk Ramsey, C., 2010 Current pretreatment methods for AMS radiocarbon dating at the Oxford Radiocarbon Accelerator Unit (ORAU), *Radiocarbon* 52, 103–112

Bronk Ramsey, C., 1995 Radiocarbon calibration and analysis of stratigraphy, *Radiocarbon* 36, 425–430

Bronk Ramsey, C., 2001 Development of the radiocarbon calibration program, *Radiocarbon* 43, 355–363

Burke, H., Phillips, E., Lee, J.R. and Wilkinson, I.P., 2009 Imbricate thrust stack model for the formation of glaciotectonic rafts: an example from the Middle Pleistocene of north Norfolk, UK, *Boreas* 38 (3), 620–637

Busschers, F.S., Kasse, C., Balen, R.T. van, Vandenberghe, J., Cohen, K.M., Weerts, H.J.T., Wallinga, J., Johns, C., Cleveringa, P. and Bunnik, F.P.M., 2007 Late Pleistocene evolution of the Rhine-Meuse system in the southern North Sea basin: imprints of climate change, sea-level oscillation and glacio-isostacy, *Quat Sci Rev* 26 (25–28), 3216–3248

Cameron, T.D.J., Crosby, A., Balson, P.S., Jeffery, D.H., Lott, G.K., Bulat, J. and Harrison, D.J., 1992 *The Geology of the Southern North Sea*, London, British Geological Survey, United Kingdom Offshore Report, HMSO

Cameron, T.D.J., Schuttenhelm, R.T.E. and Laban, C., 1989 Middle and Upper Pleistocene and Holocene stratigraphy in the southern North Sea between 52° and 54°N, 2° to 4°E, in J.P. Henriet and G. de Moor (eds), *The Quaternary and Tertiary Geology of the Southern Bight, North Sea*, Ministry of Economic Affairs, Belgian Geological Survey, 119–136

Candy, I., Silva, B. and Lee, J., 2011 Climates of the early Middle Pleistocene in Britain: environments of the earliest humans in Northern Europe, in Ashton *et al.* 2011, 11–22

Cefas, 2008 *'Best Use' – Maximising the value of data collected under the ALSF projects and the Aggregate Dredging Industry*, unpubl rep MEPF 07/08

Christensen, C., 1995 The Littorina Transgressions in Denmark, in A. Fischer (ed.), *Man and Sea in the Mesolithic*, London, Oxbow Books, 15–21

Church, J., Woodworth, P.L., Aarup, T. and Wilson, S., 2010 *Understanding Sea-level Rise and Variability*, London, Wiley-Blackwell

Clark, J.G.D., 1936 *The Mesolithic Settlement of Northern Europe*, Cambridge, Cambridge Univ Press

Clark, C.D., Hughes, A.L.C., Greenwood, S.L., Jordan, C. and Sejrup, H.P., 2012 Pattern and timing of retreat of the last British-Irish Ice Sheet, *Quat Sci Rev* 44, 112–146

Clark, P.U., Dyke, A.S., Shakun, J.D., Carlson, A.E., Clark, J., Wohlfarth, B., Mitrovica, J.X., Hostetler, S.W. and McCabe, A. M., 2009 The Last Glacial Maximum, *Science* 325 (5941), 710–4

Clark, R.L., 1982 Point count estimation of charcoal in pollen preparations and thin sections of sediments, *Pollen et Spores* 24, 523–535

Clayton, K.M., 1989 Sediment input from the Norfolk cliffs, Eastern England – a century of coast protection and its effect, *J Coastal Res* 5, 433–442

Cliquet, D. and Lautridou, J.P., 1988 Le Moustérien à Petits Bifaces Dominants de Saint-Julien de la Liegue (Eure), *Revue Archéologique de Picardie* 1–2, 175–185

Cohen, K.M., MacDonald, K., Joordens, J.C.A., Roebroeks, W. and Gibbard, P.L., 2012 The earliest occupation of north-west Europe: a coastal perspective, *Quat Int* 271, 70–83

Coles, B., 1998 Doggerland: a speculative survey, *Proc Prehist Soc* 64, 45–81

Cox, F.C., Gallois, R.W. and Wood, C.J., 1989 *Geology of the Country around Norwich*, Memoir of the British Geological Survey, Sheet 161 (England and Wales), London, HMSO

Coxon, P., 1979 *Pleistocene Environmental History in Central East Anglia*, unpubl PhD thesis for University of Cambridge

Coxon, P., 1993 The geomorphological history of the Waveney valley and the interglacial deposits at Hoxne, in R. Singer, B.G. Gladfelter, and J.J. Wymer (eds), *The Lower Paleolithic Site at Hoxne*, London, University of Chicago Press, 67–73

Cunningham, A.C. and Wallinga, J., 2012 Realizing the potential of fluvial archives using robust OSL chronologies, *Quat Geochronology* 12, 98–106

Currant, A. and Jacobi, R., 2001 A formal mammalian biostratigraphy for the Late Pleistocene of Britain, *Quat Sci Rev* 20 (16–17), 1707–1716

Daire, M., Dupont, C., Baudry, A., Billard, C., Large, J., Lespez, L., Normand, E., Scarre, C. and Bar S., (eds), 2013 *Ancient Maritime Communities and the Relationship between People and Environment along the European Atlantic Coasts*, Proceedings of the HOMER 2011 Conference, Vannes (France), Brit Archaeol Rep 2570, Oxford, Archaeopress

Darvill, T, 2008 *The Concise Oxford Dictionary of Archaeology*, Oxford, Oxford Univ Press

Dawson, A., 1984 Quaternary sea-level changes in Western Scotland, *Quat Sci Rev* 3 (4), 345–368

Debénath, A. and Dibble, H., 1994 *Handbook of Paleolithic Typology Vol 1: The Lower and Middle Paleolithic of Europe*, Philadelphia, The University Museum, University of Pennsylvania

Dellino-Musgrave, V., Gupta, S. and Russell, M., 2009 Marine aggregates and archaeology: a golden harvest? *Conservation and Management of Archaeological Sites* 11 (1), 29–42

De Loecker, D., 2010 *Great Yarmouth Dredging Licence Area 240, Norfolk, United Kingdom – Preliminary Report on the Lithic Artefacts*, unpubl typescript, Faculty of Archaeology, Leiden University, http://dx.doi.org/10.5284/1000050

De Loecker, D. and Schlanger, N., 2006 Appendix 1. Analysing Middle Palaeolithic flint assemblages: the system used for studying the flint artefacts at Maastricht-Belvédère (The Netherlands) in *Beyond the Site. The Saalian Archaeological Record at Maastricht-Belvédère (the Netherlands)*, unpubl PhD Thesis, Faculty of Archaeology, Leiden University, 303–343

Denys, L., 1992 *A check list of the diatoms in the Holocene Deposits of the Western Belgian Coastal Plain with a Survey of their Apparent Ecological Requirements: I. Introduction, ecological code and complete list*, Service Geologique de Belgique, Professional Paper No 246

Department for Communities and Local Government (DCGL), 2013 *Mineral Extraction in Great Britain 2011*, London

Dix, J. and Sturt, F., 2011 *The Relic Palaeo-landscapes of the Thames Estuary*, unpubl rep MALSF

Drees, M., 1986 Kritische kanttekeningen bij de naam "Zwarte botten fauna", *Cranium* 3, 103–120

Dutton, A. and Lambeck, K., 2012 Ice volume and sea level during the last interglacial, *Science* 337, 216–219

D'Olier, B., 2002 *Southern North Sea Sediment Transport Study Phase 2 Sediment Transport report. Appendix 10: A geological background to sediment sources, pathways and sinks*, HR Wallingford unpubl rep, Ex4526 ver 2

De Wilde, B., 2006 *Caprovis savinii* (Bovidae, Mammalia) rediscovered: core finds of an Early Pleistocene antelope from the North Sea floor, *Netherlands J Geosciences* 85, 239–243

Ehlers, J., Gibbard, P.L. and Hughes, P.D.M. (eds), 2011, *Quaternary Glaciations – Extent and Chronology A Closer Look*, Developments in Quaternary Sciences 15, London, Elsevier B.V

Emery, K., 2010 *A Re-examination of Variability in Handaxe Form in the British Palaeolithic*, unpubl PhD Thesis for University College London

EMU Ltd., 2009 *The Outer Thames Estuary Regional Environmental Characterisation*, MEPF

Erlandson, J.M., 2001 The archaeology of aquatic adaptations: paradigms for a new millennium, *J Archaeol Res* 9 (4) 287–350

Fairbanks, R.G., 1989 A 17,000-year glacio-eustatic sea-level record: influence of glacial melting rates on the Younger Dryas event and deep-ocean circulation, *Nature* 324, 637–642

Fedje, D. and Josenhans, H., 2000 Drowned forests and archaeology on the continental shelf of British Columbia, Canada, *Geology* 28, 99–102

Firth, A., 2006 Marine aggregates and prehistory, in R. Grenier, D. Nutley, and I. Cochran (eds), *Underwater Cultural Heritage at Risk: managing natural and human impacts, Heritage at Risk Special Edition*, 8–10, Paris, ICOMOS

Fischer, A., 1995 *Man and Sea in the Mesolithic*, Oxford, Oxbow Books

Fischer, A., 2004 Submerged Stone Age – Danish examples and North Sea potential, in Flemming 2004, 21–36

Fischer, A., 2011 Stone Age on the Continental Shelf: an eroding resource, in Benjamin *et al.* 2011, 298–310,

Fitch, S., Gaffney, V., Ramsey, E. and Kitchen, E., 2011 *West Coast Palaeolandcapes*, Birmingham

Flatman, J. and Doeser, J., 2010 The international management of marine aggregates and its relation to maritime archaeology, *The Historic Environment* 1 (2), 160–184

Flemming, N.C., 1969 *Archaeological Evidence for Eustatic Changes of Sea Level and Earth*

Movements in the Western Mediterranean in the Last 2000 years, Geological Society of America Special papers 109, 1–98

Flemming, N.C., 1998 *Archaeological Evidence for Vertical Movement on the Continental Shelf During the Palaeolithic, Neolithic and Bronze Age Periods,* Geological Society, London, Special Publications 146 (1), 129–146

Flemming, N.C., 2002 *The Scope of Strategic Environmental Assessment of North Sea Areas SEA 3 and SEA 2 in Regard to Prehistoric Archaeological Remains,* Technical Report SEA3_TR014, Department of Trade and Industry

Flemming, N.C., (ed.) 2004 *Submarine Prehistoric Archaeology of the North Sea: Research Priorities and Collaboration with Industry,* Counc Brit Archaeol Res Rep 141

Funnell, B.M., 1989 Quaternary, in D.G. Jenkins and J.W. Murray (eds), *Stratigraphical Atlas of Fossil Foraminifera,* London, Ellis Horwood Ltd, 563–569

Funnell, B.M., 1995 Global sea-level and the (pen-)insularity of late Cenozoic Britain, in R.C. Preece (ed.), *Island Britain: a Quaternary Perspective,* London, Geological Society Special Publication 96, 3–14

Gaffney, V., Thomson, K. and Fitch, S., 2007 *Mapping Doggerland: The Mesolithic Landscapes of the Southern North Sea,* Oxford, Archaeopress

Gamble, C., Davies, W., Pettitt, P. and Richards, M., 2004 Climate change and evolving human diversity in Europe during the last glacial, *Phil Trans Roy Soc London, B, Biological Sciences* 359 (1442), 243–53; discussion 253–4

Gibbard, P.L., 2001 History of the northwest European rivers during the past three million years, Quaternary Palaeoenvironments Group, University of Cambridge, http://www.qpg.geog.cam.ac.uk/research/projects/nweurorivers/. Accessed: January 2011

Gibbard, P.L. and Clark, C.D., 2011. Pleistocene glaciation limits in Great Britain, in Ehlers *et al.* 2011, 75–93

Gibbard, P.L., Rose, J. and Bridgland, D.R., 1988 The history of the Great Northwest European rivers during the past three million years [and discussion], *Phil Trans Roy Soc, B, Biological Sciences* 318 (1191), 559–602

Gibbard, P.L., Turner, C. and West, R.G., 2012 The Bytham River reconsidered, *Quat Int* 292, 15–32

Gijssel, K., van and van der Valk, B., 2005 Shaped by water, ice and wind: the genesis of the Netherlands, in L.P. Kooijmans, P.W. Van der Broeke, H. Frokkens, and A.L. van Gijn, (eds), *The Prehistory of the Netherlands,* Chicago, University of Chicago Press, 45–75

Glimmerveen, J., Mol, D., Post, K., Reumer, J.W.F., Van der Plicht, J., De Vos, J., Van Geel, B., Van Reenen, G. and Pals, J.P., 2004 The North Sea project. The first palaeontological, palynological and archaeological results, in Flemming (2004), 43–52

Godwin, H. and Godwin, M.E., 1933 British Maglemose harpoon sites, *Antiquity* 7 (25), 36–48

Goren-Inbar, N. and Sharon, G., 2006 *Axe Age: Acheulian Tool-making from Quarry to Discard,* London, Equinox

Gowlett, J.A.J., 2006 The early settlement of Northern Europe: fire history in the context of climate change and the social brain, *Comptes Rendus Palevol* 5, 299–310

Graham, A.G.C., Stoker, M.S., Lonergan, L., Bradwell, T. and Stewart, M.A., 2011 The Pleistocene glaciations of the North Sea basin, in Ehlers *et al.* 2011, 261–278

Grün, R. and Schwarcz, H.P., 2000 Revised open system U-series/ESR age calculations for teeth from Stratum C at the Hoxnian Interglacial type locality, England, *Quat Sci Rev* 19 (12), 1151–1154

Gupta, S., Collier, J.S.J., Palmer-Felgate, A. and Potter, G., 2007 Catastrophic flooding origin of shelf valley systems in the English Channel, *Nature* 448 (7151), 342–5

Hamel, A., 2011 *Wrecks at Sea: Research in Context,* London, MEPF

Harding, P., Bridgland, D.R., Allen, P., Bradley, P., Grant, M.J., Peat, D., Schwenninger, J.-L., Scott, R., Westaway, R., and White, T.S., 2012 Chronology of the Lower and Middle Palaeolithic in NW Europe: developer-funded investigations at Dunbridge, Hampshire, southern England, *Proc Geol Ass* 123 (4), 484–607

Hartley, B., Barber, H.G., Carter, J.R. and Sims, P.A., 1996 *An Atlas of British Diatoms,* Bristol, Biopress Limited

Hendley, N.I., 1964 *An Introductory Account of the Smaller Algae of British Coastal Waters. Part V. Bacillariophyceae (Diatoms),* Ministry of Agriculture Fisheries and Food, Series IV, London, HMSO

Higham, T., 2011 European Middle and Upper Palaeolithic radiocarbon dates are often older than they look: problems with previous dates and some remedies, *Antiquity* 85, 235–249

Highley, D.E., Hetherington, L.E., Brown, T.J., Harrison, D.J. and Jenkins, G.O., 2007 *The strategic importance of the marine aggregate industry to the United Kingdom,* Research report OR/07/019, London, British Geological Survey

Hijma, M.P., Cohen, K.M., Roebroeks, W., Westerhoff, W.E. and Busschers, F.S., 2012 Pleistocene Rhine-Thames landscapes: geological background for hominin occupation of the southern North Sea region, *J Quat Sci* 27 (1), 17–39

Hill, T., Fletcher, W. and Good, C., 2008 *The Suffolk Valleys River Project: a review of published and grey archaeological and palaeoenvironmental literature*, The Suffolk Valleys River Project ALSF 4772, Birmingham, Birmingham University

Hodder, I., 1986 *Reading the Past: Current Approaches to Interpretation in Archaeology*, Cambridge, Cambridge Univ Press

Hodgson, J.M., 1976 *Soil Survey Field Handbook: Describing and Sampling Soil Profiles*, Technical Monogr. – Soil Survey 5, Harpenden, Rothamsted Experimental Station, Lawes Agricultural Trust

Hosfield, R., 1999 *The Palaeolithic of the Hampshire Basin*, Brit Archaeol Rep 286, Oxford, Archaeopress

Hosfield, R., 2007 Terrestrial implications for the maritime geoarchaeological resource: a view from the Lower Palaeolithic, *J Maritime Archaeol* 2 (1), 4–23

Hosfield, R., 2011 The British Lower Palaeolithic of the early Middle Pleistocene, *Quat Sci Rev*, 30 (11–12), 1486–1510

Hosfield, R. and Chambers, J., 2004 *The Archaeological Potential of Secondary Contexts*, Southampton, University of Southampton

Housley, R.A., Gamble, C.S., Street, M. and Pettitt, P., 1997 Radiocarbon evidence for the late glacial human recolonisation of northern Europe, *Proc Prehist Soc* 63, 25–54

Hublin, J.-J., Weston, D., Gunz, P., Richards, M., Roebroeks, W., Glimmerveen, J. and Anthonis, L., 2009 Out of the North Sea: the Zeeland ridges Neandertal, *J Hum Evol* 57 (6), 777–85

Hustedt, F., 1953 Die Systematik der Diatomeen in ihren Beziehungen zur Geologie und Okologie nebst einer Revision des Halobien-systems, *Sv. Bot. Tidskr* 47, 509–519

Hustedt, F., 1957 Die Diatomeenflora des Flusssystems der Weser im Gebiet der Hansestadt Bremen, *Ab. naturw. Ver. Bremen* 34, 181–440

Huuse, M. and Lykke-Andersen, H., 2000 Overdeepened Quaternary valleys in the eastern Danish North Sea: morphology and origin, *Quat Sci Rev* 19 (12), 1233–1253

Jacobi, R. and Higham, T., 2011 The Later Upper Palaeolithic recolonisation of Britain: new results from AMS Radiocarbon Dating, in Ashton *et al.* 2011, 223–247

James, J.W.C., Pearce, B., Coggan, R.A., Arnott, S.H.L., Clark, R., Plim, J.F., Pinnion, J., Barrio Frójan, C., Gardiner, J.P., Morando, A., Baggaley, P.A., Scott, G. and Bigourdan, N., 2010 *The South Coast Regional Environmental Characterisation*, British Geological Survey Open Report OR/09/51

James, J.W.C., Pearce, B., Coggan, R.A., Leivers, M., Clark, R.W.E., Plim, J.F., Hill, J.M., Arnott, S.H.L., Bateson, L., De-Burgh Thomas, A. and Baggaley, P.A., 2011 *The MALSF Synthesis Study in the Central and Eastern English Channel*, British Geological Survey Open Report OR/11/01

Jarvis, A., Reuter, H.I., Nelson, A. and Guevara, E., 2008 Hole-filled seamless SRTM data V4, *International Centre for Tropical Agriculture (CIAT)*

Johnson, L.L. and Stright, M., 1992 *Paleoshorelines and Prehistory: An Investigation of Method*, Boca Raton, CRC Press

Kerney, M.P., 1999 *Atlas of the Land and Freshwater Molluscs of Britain and Ireland*, Colchester, Harley Books

Kolen, J., De Loecker, D., Groenendijk, A. and De Warrimont, J.-P., 1999 Middle Palaeolithic surface scatters: how informative? A case study from southern Limburg (The Netherlands), in W. Roebroeks, and C. Gamble, (eds), *The Middle Palaeolithic Occupation of Europe*, Leiden, University of Leiden, 177–191

Kopp, R.E., Simons, F.J., Mitrovica, J.X., Maloof, A.C. and Oppenheimer, M., 2009 Probabilistic assessment of sea level during the last interglacial stage, *Nature* 462, 863–867

Krammer, K. and Lange-Bertalot, H., 1986–1991 *Bacillariophyceae*, Stuttgart, Gustav Fisher Verlag

Laban, C. and van der Meer, J.J.M., 2011 Pleistocene Glaciation in The Netherlands, in Ehlers *et al.* 2011, 247–260

Lambeck, K., Smither, C. and Johnston, P., 1998 Sea-level change, glacial rebound and mantle viscosity for northern Europe, *Geophysical J Int* 134, 102–144

Lambeck, K., Yokoyama, Y., Johnston, P. and Purcell, A., 2001. Corrigendum to "Global ice volumes at the Last Glacial Maximum and early Late glacial", *Earth and Planetary Science Letters* 190, 275

Lankelma Andrews, 2007 *Areas 240, 242, 328A and 328B. Vibrocore Survey*, unpubl rep

Leary, J., 2011 Experiencing change on the prehistoric shores of Northsealand: an anthropological perspective on Early Holocene sea-level rise, in Benjamin *et al.* 2011, 75–84

Lee, J.R., 2009 Patterns of preglacial sedimentation and glaciotectonic deformation within early Middle Pleistocene sediments at Sidestrand, north Norfolk, UK, *Proc Geol Ass* 120 (1), 34–48

Lee, J.R., Rose, J., Candy, I. and Barendregt, R.W., 2006 Sea-level changes, river activity, soil development and glaciation around the western margins of the southern North Sea Basin during the Early and early Middle Pleistocene: evidence from Pakefield, Suffolk, UK, *J Quat Sci* 21 (2), 155–179

Lee, J.R., Rose, J., Hamblin, R.J. and Moorlock, B.S., 2004 Dating the earliest lowland glaciation of eastern England: a pre-MIS 12 early Middle Pleistocene Happisburgh glaciation, *Quat Sci Rev* 23 (14–15), 1551–1566

Lee, J.R., Rose, J., Hamblin, R.J.O., Moorlock, B.S.P., Riding, J.B., Phillips, E., Barenbregt, R.W. and Candy, I., 2011 The glacial history of the British Isles during the Early and Middle Pleistocene: Implications for the long-term development of the British Ice Sheet, in Ehlers *et al.* 2011, 59–74

Leroy, S.A.G., Arpe, K. and Mikolajewicz, U., 2011 Vegetation context and climatic limits of the Early Pleistocene hominin dispersal in Europe, *Quat Sci Rev* 30 (11), 1448–1463

Lewin, J. and Gibbard, P.L., 2010 Quaternary river terraces in England: forms, sediments and processes, *Geomorphology* 120 (3–4), 293–311

Lewis, S.G., Ashton, N. and Jacobi, R., 2011 Testing human presence during the Last Interglacial (MIS 5e): a review of the British evidence, in Ashton *et al.* 2011, 125–164

Limpenny, S.E., Barrio Froján, C., Cotterill, C., Foster-Smith, R.L., Pearce, B., Tizzard, L., Limpenny, D.L., Long, D., Walmsley, S., Kirby, S., Baker, K., Meadows, W.J., Rees, J., Hill, J., Wilson, C., Leivers, M., Churchley, S., Russell, J., Birchenough, A.C., Green, S.L. and Law, R.J., 2011 *The East Coast Regional Environmental Characterisation*, Cefas Open, MEPF

Lisiecki, L.E. and Raymo, M.E., 2005 A Pliocene-Pleistocene stack of 57 globally distributed benthic $\delta 18O$ records, *Paleoceanography* 20 (1), 1–17

Long, D., Wickham-Jones, C.R. and Ruckley, N.A., 1986 A flint artefact from the northern North Sea, in D.A. Roe (ed), *Studies in the Upper Palaeolithic of Britain and Northwest Europe*, Oxford, Brit Archaeol Rep 296, 55–62

Lowe, J.J. and Walker, M.J.C., 1997 *Reconstructing Quaternary Environments*, 2nd edn Essex, Pearson Prentice Hall

Lübke, H., Schmolcke, U. and Tauber, F., 2011 Mesolithic hunter-fishers in a changing world: a case study of submerged sites on the Jackelberg, Wismar Bay, northeastern Germany, in Benjamin *et al.* 2011, 21–37

Lurton, X., 2002 *An Introduction to Underwater Acoustics: Principles and Applications*, London, Springer

Masters, P. and Flemming, N.C., 1983 *Quaternary Coastlines and Marine Archaeology*, London, Academic Press Limited

McManamon, F.P., 1984 Discovering sites unseen, *Advances in Archaeological Method and Theory*, 7, 223–292

McNabb, J., 1992 *The Clactonian: British Lower Palaeolithic flint technology in biface and non-biface assemblages*, London, PhD dissertation, University of London

Meisch, C., 2000 *Freshwater Ostracoda of Western and Central Europe*, Suesswasserfauna von Mitteleuropa 8/3, Heidelberg, Spektrum Akademischer Verlag

Mellett, C.L., Hodgson, D.M., Lang, A., Mauz, B., Selby, I. and Plater, A.J., 2012 Preservation of a drowned gravel barrier complex: a landscape evolution study from the north-eastern English Channel, *Marine Geology* 315–318, 115–131

Mol, D., Post, K., Reumer, J.W.F., Van der Plicht, J., De Vos, J., Van Geel, B., Van Reenen, G., Pals, J.P. and Glimmerveen, J., 2006 The Eurogeul – first report of the palaeontological, palynological and archaeological investigations of this part of the North Sea, *Quat Int* 142–143, 178–185

Momber, G., Tomalin, D., Scaife, R., Satchell, J. and Gillespie, J., 2011 *Mesolithic Occupation at Bouldner Cliff and the Submerged Prehistory Landscapes of the Solent*, Counc Brit Archaeol Rep 164, York

Mook, W.G., 1986 Business meeting: recommendations/resolutions adopted by the Twelfth International Radiocarbon Conference, *Radiocarbon* 28, 799

Moore, P.D., Webb, J.A. and Collinson, M.E., 1991 *Pollen Analysis*, 2nd edn, Oxford, Blackwell Scientific Publications

Moorlock, B.S.P., Hamblin, R.J.O., Booth, S.J. and Morigi, A.N., 2000 *Geology of the Country around Lowestoft and Saxmundham*, London, HMSO

Murray, J.W., 1979 *British Nearshore Foraminiferids*, London, Academic Press

Murray, J.W., 1991 *Ecology and Palaeoecology of Benthic Foraminifera*, Essex, Longman Scientific

Murton, D.K. and Murton, J.B., 2012 Middle and Late Pleistocene glacial lakes of lowland Britain and the southern North Sea Basin, *Quat Int* 260, 115–142

Newell, R.C., Woodcock, T.A. (eds), 2013 *Aggregate Dredging and the Marine Environment: an overview of recent research and current industry practice*, The Crown Estate

Nickens, P.R., 1991 The destruction of archaeological sites and data in G.E. Smith and J.E. Ehrenhard (eds), *Protecting the Past*, Boca Raton, CRC Press, 73–81

Oxford Archaeology, 2007 *England's Historic Seascapes Pilot Study: Southwold-to-Clacton final project report*, unpubl rep for English Heritage

Parfitt, S. A, Ashton, N.M., Lewis, S.G., Abel, R.L., Coope, G.R., Field, M.H., Gale, R., Hoare, P.G., Larkin, N.R., Lewis, M.D., Karloukovski, V., Maher, B. A, Peglar, S.M., Preece, R.C., Whittaker, J.E. and Stringer, C.B., 2010 Early Pleistocene human occupation at the edge of the boreal zone in northwest Europe, *Nature* 466 (7303), 229–33

Parfitt, S. A, Barendregt, R.W., Breda, M., Candy, I., Collins, M.J., Coope, G.R., Durbidge, P., Field, M.H., Lee, J.R., Lister, A.M., Mutch, R., Penkman, K.E.H., Preece, R.C., Rose, J., Stringer, C.B., Symmons, R., Whittaker, J.E., Wymer, J.J. and Stuart, A.J., 2005. The earliest record of human activity in northern Europe, *Nature* 438 (7070), 1008–12

Parkinson, R., 2001 *High Resolution Site Surveys*, London, Taylor and Francis Group

Pawley, S.M., Bailey, R.M., Rose, J., Moorlock, B.S.P., Hamblin, R.J.O., Booth, S.J. and Lee, J.R., 2008 Age limits on Middle Pleistocene glacial sediments from OSL dating, north Norfolk, UK, *Quat Sci Rev* 27 (13–14), 1363–1377

Peeters, H., Murphy, P. and Flemming, N.C. (eds), 2009 *North Sea Prehistory Research and Management Framework (NSPRMF)*, Rijksdienst voor het Cultureel Erfgoed and English Heritage, Amersfoort, The Netherlands

Penkman, K., Collins, M., Keen, D. and Preece, R., 2008 *British Aggregates: An Improved Chronology Using Amino Acid Racemization and Degradtion of Intra-crystalline Amino Acids (IcPD)*, Scientific Dating Report No. 6–2008, ALSF, English Heritage

Pettitt, P., Gamble, C. and Last, J. (eds), 2008 *Research and Conservation Framework for the British Palaeolithic*, London, The Prehistoric Society and English Heritage

Pettitt, P. and White, M.J., 2012 *The British Palaeolithic: Human Societies at the Edge of the Pleistocene World*, Abingdon, Routledge

Pirazzoli, P.A., 1985 Sea-level change, *Nature and Resources* 21, 2–9

Pirazzoli, P.A., 1996 *Sea-level Changes: The Last 20,000 years*, Chichester, Wiley

Pope, M., 2010 *Valdoe Assessment Survey*, unpubl rep ALSF Project Number 4620, English Heritage

Preece, R.C. and Parfitt., S.A., 2008 The Cromer Forest-bed Formation: some recent developments relating to early human occupation and lowland glaciation, in I. Candy, J.R. Lee, and A.M. Harrison (eds), *The Quaternary of Northern East Anglia. Field Guide*, London, Quaternary Research Association, 60–83

Ransley, J., Sturt, F., Dix, J., Adams, J. and Blue, L., (eds), 2013 *People and the Sea: A Maritime Archaeological Research Agenda for England*, York, Counc Brit Archaeol Res Rep 171

Read, A., Godwin, M., Mills, C. A., Juby, C., Lee, J.R., Palmer, A.P., Candy, I. and Rose, J., 2007 Evidence for Middle Pleistocene temperate-climate high sea-level and lowland-scale glaciation, Chapel Hill, Norwich, UK, *Proc Geol Ass* 118 (2), 143–156

Reid, C., 1913 *Submerged Forests*, Cambridge, Cambridge Univ Press

Reimer, P.J., Baillie, M.G.L., Bard, E., Bayliss, A., Beck, J.W., Blackwell, P.G., Bronk Ramsey, C., Buck, C.E., Burr, G.S., Edwards, R.L., Friedrich, M., Grootes, P.M., Guilderson, T.P., Hajdas, I., Heaton, T.J., Hogg, A.G., Hughen, K.A., Kaiser, K.F., Kromer, B., McCormac, G., Manning, S., Reimer, R.W., Remmele, S., Richards, D.A., Southon, J.R., Talamo, S., Taylor, F.W., Turney, C. M., Van der Plicht, J. and Weyhenmeyer, C.E., 2009 INTCAL09 and MARINE09 radiocarbon age calibration curves, 0–50,000 years cal BP, *Radiocarbon* 41 (4), 1111–1150

Rendell, H.M., 1995 *Luminescence Dating of Quaternary Sediments*, Geological Society, London, Special Publications 89 (1), 223–235

Rensink, E. and Stapert, D., 2005 The first "modern" humans Upper Paleolithic, in L.P. Kooijams, P.W. van den Broeke, H. Fokkens, and A.L. van Gign (eds), *The Prehistory of the Netherlands*, Chicago, University of Chicago Press, 115–134

Roberts, M.B. and Parfitt, S., 1999 *Boxgrove: A Middle Pleistocene hominid site at Eartham Quarry, Boxgrove, West Sussex*, London, English Heritage, Archaeol Rep 17

Rodwell, J.S. (ed.), 1991 *British Plant Communities. Volume 1. Woodlands and Scrub*, Cambridge, Cambridge, Univ Press

Roe, D., 1967 *A Study of Handaxe Groups of the British Lower and Middle Palaeolithic Periods, Using Methods of Metrical and Statistical Analysis, with a Gazetteer of British Lower and Middle Palaeolithic Sites*, unpubl PhD thesis for University of Cambridge

Roe, D.A., 1968 British Lower and Middle Palaeolithic handaxe group, *Proc Prehist Soc* 34, 1–82

Roebroeks, W., Hublin, J. and Macdonald, K., 2011 Continuities and discontinuities in Neandertal presence: a closer look at Northwestern Europe, in Ashton *et al.* 2011, 113–123

Rose, J., 2009 Early and Middle Pleistocene landscapes of eastern England, *Proc Geol Ass* 120 (1), 3–33

Rose, J., Candy, I., Moorlock, B.S.P., Wilkins, H., Lee, J.R.J. A., Hamblin, R.J.O., Riding, J.B. and Morigi, A. N., 2002 Early and early Middle Pleistocene river, coastal and neotectonic processes, southeast Norfolk, England, *Proc Geol Ass* 113 (1), 47–67

Rose, J., Moorlock, B.B.S.P. and Hamblin, R.J.O., 2001 Pre-Anglian fluvial and coastal deposits in Eastern England: lithostratigraphy and palaeoenvironments, *Quat Int* 79 (1), 5–22

Ruebens, K., 2006 A typological dilemma: Micquian elements in continental northwestern Europe during the last glacial cycle (MIS 5d–3), *Lithics* 27, 58–73

Russell, M. and Firth, A., 2007 *Working alongside the marine Historic Environment – An Aggregate Dredging Industry Perspective*, CEDA Dredging Day publication

Sainty, J.E., 1927 An Acheulian Palaeolithic workshop site at Whitlingham, near Norwich. With geological notes on the Acheulian site at Whitlingham, Norfolk by Professor P.G.H. Boswell, *Proc Prehist Soc of E Anglia* 5, 177–213

Sainty, J.E., 1933 Some Norfolk Palaeolithic discoveries. With an appendix on Implementiferous gravels in East Anglia by Dr J.D. Solomon, *Proc Prehist Soc of E Anglia* 7 (2), 171–176

Sanderson, D.C.W. and Murphy, S., 2010 Using simple portable OSL measurements and laboratory characterisation to help understand complex and heterogeneous sediment sequences for luminescence dating, *Quat Geochronology* 5 (2–3), 299–305

Santonja, M. and Villa, P., 2006 The Acheulian of Western Europe, in N. Goren-Inbar and G. Sharon (eds), *Axe Age: Acheulian Tool-making from Quarry to Discard*, London, Equinox, 429–478

Schiffer, M.B., 1976 *Behavioral Archaeology*, London, Academic Press

Schiffer, M.B., 1983 Toward the identification of formation processes, *American Antiquity* 48 (4), 675–705

Schiffer, M.B., 1987 *Formation Processes of the Archaeological Record*, Salt Lake City, University of Utah Press

Schiffer, M.B., Sullivan, A.P. and Klinger, T.C., 1978 The design of archaeological surveys, *World Archaeology* 10, 1–28

Schreve, D. (ed.), 2004 *The Quaternary Mammals of Southern and Eastern England*, London, Quaternary Research Association

Schreve, D.C., Bridgland, D.R., Allen, P., Blackford, J.J., Gleed-Owen, C.P., Griffiths, H.I., Keen, D.H. and White, M.J., 2002 Sedimentology, palaeontology and archaeology of late Middle Pleistocene River Thames terrace deposits at Purfleet, Essex, UK, *Quat Sci Rev* 21 (12–13), 1423–1464

Schreve, D., Harding, P. and White, M., 2006 A Levallois knapping site at West Thurrock, Lower Thames, UK: its quaternary context, environment and age, *Proc Prehist Soc* 72, 21–52

Scott, B. and Ashton, N., 2011 The Early Middle Palaeolithic: The European Context, in Ashton *et al.* 2011, 91–112

Scott, E.M., 2003 The third international radiocarbon intercomparison (TIRI) and the fourth international radiocarbon inter-comparison (FIRI) 1990–2002: results, analyses, and conclusions, *Radiocarbon* 45, 135–408

Scuvée, F. and Verague, J., 1988 *Le gisement sous-marin du Paléolithique moyen de l'anse de la mondrée à Fermanville, Manche*, Cherbourg, C.E.H.P-Littus

Shackleton, J.C., Andel, T.H. van and Runnels, C., 1984 Coastal paleogeography of the Central and Western Mediterranean during the last 125,000 years and its archaeological implications, *J Fld Archaeol* 11 (3), 307–314

Shelley, P.H., 1990 Variation in lithic assemblages: an experiment, *J Fld Archaeol* 17, 187–193

Shennan, I. and Horton, B., 2002 Holocene land- and sea-level changes in Great Britain, *J Quat Sci* 17 (5–6), 511–526

Shennan, I., Lambeck, K., Horton, B., Innes, J., Lloyd, J., McArthur, J., Purcell, T. and Rutherford, M., 2000 Late Devensian and Holocene records of relative sea-level changes in northwest Scotland and their implications for glacio-hydro-isostatic modelling, *Quat Sci Rev* 19 (11), 1103–1135

Shennan, I., Bradley, S., Milne, G., Brooks, A., Bassett, S. and Hamilton, S., 2006 Relative sea-level changes, glacial isostatic modelling and ice-sheet reconstructions from the British Isles since the Last Glacial Maximum, *J Quat Sci* 21, 585–599

Shennan, I., Milne, G. and Bradley, S., 2012 Late Holocene vertical land motion and relative sea-level changes: lessons from the British Isles, *J Quat Sci* 27 (1), 64–70

Sheriff, R.E. and Geldart, L.P., 1983 *Exploration Seismology*, Cambridge, Cambridge Univ Press

Shipman, P., 1981 *Life History of a Fossil: An Introduction to Taphonomy and Paleoecology*, Cambridge Massachusetts, Harvard Univ Press

Sidall, M., Rohling, J., Almogi-Labin, A., Hemleben, C., Meischner, D., Schmelzer, I., Smeed, D.A., 2003 Sea-level fluctuations during the last glacial cycle, *Nature* 423, 19–24

Singer, R., Gladfelter, B.G. and Wymer, J.J. (eds), 1993 *The Lower Paleolithic Site at Hoxne, England*, London, University of Chicago Press

Slota Jr, P.J., Jull, A.J.T., Linick, T.W. and Toolin, L.J., 1987 Preparation of small samples for 14C accelerator targets by catalytic reduction of CO, *Radiocarbon* 29, 303–306

Soressi, M., 2002 *Le Moustérien de Tradition Acheuléene du Sud-Ouest de la France*, unpubl PhD Thesis, University of Bordeaux

Stace, C., 1997 *New Flora of the British Isles*, 2nd edn, Cambridge, Cambridge Univ Press

Stapert, D., 1976 Some natural surface modifications on flint in the Netherlands, *Prehistoria* 18, 7–41

Stokes, S., Ingram, S., Aitken, M.J. and Sirocko, F., 2003 Alternative chronologies for Late Quaternary (Last Interglacial-Holocene) deep sea sediments via optical dating of silt-sized quartz, *Quat Sci* 22, 925–941

Strijdonk, H., Post, K., Mol, D. and Ras, B., 2011 *Report of Identifications of Fossil Mammal Remains which have been Collected in Sediments at the Premises of SBV (Sorteerbedrijf Vlissingen) dredged from the North Sea bottom off the coast of East Anglia*, unpubl rep

Strijdonk, H., Post, K., Mol, D. and Ras, B., 2012 *Report 2 of Identifications of Fossil Mammal Remains which have been Collected in Sediments at the Premises of SBV (Sorteerbedrijf Vlissingen) dredged from the North Sea bottom off the coast of East Anglia*, unpubl rep

Stuiver, M. and Kra, R.S., 1986 Editorial Comment, *Radiocarbon* 2B, ii

Stuiver, M. and Polach, H., 1977 Discussion: Reporting of ^{14}C data, *Radiocarbon* 19 (3), 355–63

Sturt, F. and Standen, T., 2013 *NHPP 3A1: Unknown Marine Assets and Landscapes. The Social Context of Submerged Prehistoric Landscapes*, Southampton, Univ Southampton

Sumbler, M.G. (eds), 1996 *British Regional Geology: London and the Thames Valley*. 4th edn, London, HMSO for the British Geological Survey

Tappin, D.R., Pearce, B., Fitch, S., Dove, D., Gearey, B., Hill, J.M., Chambers, C., Bates, R., Pinnion, J., Diaz Doce, D., Green, M., Gallyot, J., Georgiou, L., Brutto, D., Marzialetti, S., Hopla, E., Ramsay, E. and Fielding, H., 2011 *The Humber Regional Environmental Characterisation*, British Geological Survey Open Report OR/10/54

The Crown Estate and BMAPA, 2009 *Marine Aggregate Dredging Ten Year Review. The Area Involved 1998–2008*, The Crown Estate and BMAPA

The Crown Estate and BMAPA, 2010 *Marine Aggregate Terminology: A Glossary*, The Crown Estate and BMAPA

The Crown Estate and BMAPA, 2013 *The Area involved – the 15th Annual Report. Marine Aggregate Dredging 2012*, The Crown Estate and BMAPA

Tizzard, L., Baggaley, P.A., and Firth, A.J., 2011 Seabed Prehistory: Investigating Palaeolandsurfaces with Palaeolithic Remains from the Southern North Sea, in Benjamin *et al.* 2011, 65–74

Tizzard, L., Bicket, A.R., Benjamin, B., De Loecker, D., 2014 A Middle Palaeolithic site in the southern North Sea: investigating the archaeology and palaeogeography of Area 240, *J Quat Sci* 29, DOI: 10.1002/jqs.2743

Tolan-Smith, C., 2008 Mesolithic Britain, in G. Bailey and P. Spikins (eds), *Mesolithic Europe*, Cambridge, Cambridge Univ Press, 132–157

Toms, P., 2011 *Seabed Prehistory Area 240: Optical Dating of Submarine Cores Scientific Dating Report*, Research Department Report Series no. 81-2011, English Heritage

Toucanne, S., Zaragosi, S., Bourillet, J.F.F., Gibbard, P.L.L., Eynaud, F., Giraudeau, J., Turon, J.L., Cremer, M., Cortijo, E., Martinez, P. and Rossignol, L., 2009 A 1.2Ma record of glaciation and fluvial discharge from the West European Atlantic margin, *Quat Sci Rev* 28 (25–26), 2974–2981

Tzedakis, P.C., 2005 Towards an understanding of the response of southern European vegetation to orbital and suborbital climate variability, *Quat Sci Rev* 24, 1585–1599

Uldum, O., 2011 The excavation of a Mesolithic double burial from Tybrind Vig, Denmark, in Benjamin *et al.* 2011, 15–20

Van Kolfschoten, T. and Laban, C., 1995. Pleistocene terrestrial mammal faunas from the North Sea, *Mededelingen Rijks Geologische Dienst* 52, 135–151

Vandeputte, K., Moens, L. and Dams, R., 1996. Improved sealed-tube combustion of organic samples to CO_2 for stable isotope analysis, radiocarbon dating and percent carbon determinations, *Analytical Letters* 29 (15), 2761–2763

Van Peer, P., 1992 *The Levallois Reduction Strategy*, Madison Wisconsin, Monographs in World Archaeology 13

Vos, P.C. and de Wolf, H., 1988 Methodological aspects of palaeoecological diatom research in coastal areas of the Netherlands, *Geologie en Mijnbouw* 67, 31–40

Vos, P.C. and de Wolf, H., 1993 Diatoms as a tool for reconstructing sedimentary environments in coastal wetlands; methodological aspects, *Hydrobiologia* 269/270, 285–296

Waelbroeck, C., Labeyrie, L., Michel, E., Duplessy, J.C., McManus, J.F., Lambeck, K., Balbon, E. and Labracherie, M., 2002 Sea-level and deep water temperature changes derived from benthic foraminifera isotopic records, *Quat Sci Rev* 21 (1–3), 295–305

Weerts, H., Otte, A., Smit, B. and Vos, P., 2012 Finding the needle in the haystack by using knowledge of Mesolithic human adaptation in a drowning delta, *eTopoi Journal for Ancient Studies* 3, 17–24

Wenban-Smith, F.F., 2002 *Marine Aggregate Dredging and the Historic Environment: Palaeolithic and Mesolithic archaeology on the seabed*, London, BMAPA and English Heritage

Wenban-Smith, F.F., 2004 Handaxe typology and Lower Palaeolithic cultural development: ficrons, cleavers and two giant handaxes from Cuxton, in M. Pope and K. Cramp (eds), *Lithics* 25 (Papers in Honour of R.J. MacRae), 11–21

Wenban-Smith, F.F., (ed.) 2013 *The Ebbsfleet Elephant: Excavations at Southfleet Road, Swanscombe in advance of High Speed 1, 2003–4*, Oxford Archaeology Monogr. 20

Wenban-Smith, F.F., Allen, P., Bates, M.R., Parfitt, S.A., Preece, R.C., Stewart, J.R., Turner, C. and Whittaker, J.E., 2006 The Clactonian elephant butchery site at Southfleet Road, Ebbsfleet, UK, *J Quat Sci* 21 (5), 471–483

Wenban-Smith, F.F., Bates, M.R. and Schwenninger, J.-L., 2010 Early Devensian (MIS 5d-5b) occupation at Dartford, southeast England, *J Quat Sci* 25 (8), 1193–1199

Wenban-Smith, F.F. and Bridgland, D.R., 2001 Palaeolithic archaeology at the Swan Valley Community School, Swanscombe, Kent, *Proc Prehist Soc* 66, 209–255

Wenban-Smith, F.F., Gamble, C.S. and ApSimon, A.M., 2000 The Lower Palaeolithic site at Red Barns, Portchester, Hampshire: bifacial technology, raw material quality and the organisation of Archaic behaviour, *Proc Prehist Soc* 66, 209–255

Werff, A., van der and Huls., H., 1957–1974 *Diatomeenflora van Nederland*, Koenigstein, Germany, Otto Koetlz Science Publishers

Wessex Archaeology, 2003 *Artefacts from the Sea. Catalogue of the Michael White Collection*, ref 51541a and b http://dx.doi.org/10.5284/1000260

Wessex Archaeology, 2008a *Seabed in Prehistory: Gauging the Effects of Marine Aggregate Dredging. Final Report Volume II Arun*, ref 57422.32 http://dx.doi.org/10.5284/1000050

Wessex Archaeology, 2008b *Seabed in Prehistory: Gauging the Effects of Marine Aggregate Dredging. Final Report Volume III Arun Additional Grabbing*, ref 57422.33 http://dx.doi.org/10.5284/1000050

Wessex Archaeology, 2008c *Seabed in Prehistory: Gauging the Effects of Marine Aggregate Dredging. Final Report Volume IV Great Yarmouth*, ref 57422.34 http://dx.doi.org/10.5284/1000050

Wessex Archaeology, 2008d *Seabed in Prehistory: Gauging the Effects of Marine Aggregate Dredging. Final Report Volume V Eastern English Channel*, ref 57422.35 http://dx.doi.org/10.5284/1000050

Wessex Archaeology, 2008e *Seabed in Prehistory: Gauging the Effects of Marine Aggregate Dredging. Final Report Volume VI Humber*, ref 57422.36 http://dx.doi.org/10.5284/1000050

Wessex Archaeology, 2008f *Seabed in Prehistory: Gauging the Effects of Marine Aggregate Dredging. Final Report Volume VII Happisburgh and Pakefield Exposures*, ref 57422.37 http://dx.doi.org/10.5284/1000050

Wessex Archaeology, 2008g *Aggregate Levy Sustainability Fund, English Heritage, Round 2 Continuation. Seabed Grab Sampling Seminar*, ref 65700.01

Wessex Archaeology, 2009a *Seabed Prehistory: Site Evaluation Techniques (Area 240): Existing Data Review*, ref 70751.03 http://dx.doi.org/10.5284/1000050

Wessex Archaeology, 2009b *Seabed Prehistory: Site Evaluation Techniques (Area 240): Geophysical Survey*, ref 70751.04 http://dx.doi.org/10.5284/1000050

Wessex Archaeology, 2009c *Seabed Prehistory: Site Evaluation Techniques (Area 240): Palaeo-Environmental Sampling*, ref 70753.02

Wessex Archaeology, 2010a *Seabed Prehistory: Site Evaluation Techniques (Area 240): Palaeo-Environmental Sampling*, ref 70753.02. http://dx.doi.org/10.5284/1000050

Wessex Archaeology, 2010b *Seabed Prehistory: Site Evaluation Techniques (Area 240): Seabed Sampling*, ref 70752.02 http://dx.doi.org/10.5284/1000050

Wessex Archaeology, 2011 *Seabed Prehistory: Site Evaluation Techniques (Area 240): Synthesis*, ref 70753.02 http://dx.doi.org/10.5284/1000050

Wessex Archaeology, 2012a *Licence Area 240 Archaeological Monitoring of Dredging Activity*, ref 77860.02

Wessex Archaeology, 2012b *Licence Area 240 Archaeological Mitigation: Frindsbury Wharf Methodological Trial*, ref 77860.03

Wessex Archaeology, 2013a *Audit of current state of knowledge of submerged palaeolandscapes and sites*, ref 84570.01 http://www.english-heritage.org.uk/publications/audit-current-state-knowledge-submerged-palaeolandscapes-sites

Wessex Archaeology, 2013b *Palaeo-Yare Catchment Assessment. Technical Report*, unpubl rep 83740.04

Westaway, R., 2009 Quaternary vertical crustal motion and drainage evolution in East Anglia and adjoining parts of southern England: chronology of the Ingham River terrace deposits, *Boreas* 38 (2), 261–284

Westley, K., Dix, J. and Quinn, R., 2004 *Re-assessment of the Archaeological Potential of Continental Shelves*, unpubl rep for English Heritage ALSF project no. 3362. School of Ocean and Earth Science, University of Southampton

White, M.J., 1998a. On the significance of Acheulian biface variability in Southern Britain. *Proc Prehist Soc* 64, 15–44

White, M.J., 1998b Twisted ovate bifaces in the British Lower Palaeolithic: some observations and implications, in N. Ashton, F. Healy, and P. Pettitt (eds), *Stone Age Archaeology: Essays in Honour of John Wymer*, Oxford, Oxbow Books, 98–104

White, M.J., 2006 Things to do in Doggerland when you're dead: surviving OIS3 at the northwestern-most fringe of Middle Palaeolithic Europe, *World Archaeology* 38(4), 547–75

White, M.J. and Ashton, N.M., 2003 Lower Palaeolithic core technology and the origins of the Levallois method in NW Europe, *Current Anthropology* 44, 598–609

White, M.J. and Jacobi, R.M., 2002 Two Sides to Every Story: Bout Coupé Handaxes Revisited, *Oxford J Archaeol* 21, 109–133

White, M.J., Scott, B., and Ashton, N.M., 2006 The Early Middle Palaeolithic, in Britain: archaeology, settlement history and human behaviour, *J Quat Sci* 21 (5), 525–541

Woodcock, A.G., 1978 The Palaeolithic of Sussex, in P.L. Drewett (ed.), *Archaeology in Sussex to AD 1500*, Counc Brit Archaeol Res Rep 29, 8–14

Wragg-Sykes, R.M., 2009 *The British Mousterian: Late Neanderthal archaeology in landscape context*, unpubl PhD thesis, University of Sheffield

Wymer, J.J., 1968 *Lower Palaeolithic Archaeology in Britain, as Represented by the Thames*, London, John Baker

Wymer, J.J., 1985 *The Palaeolithic Sites of East Anglia*, Norwich, Geo Books

Wymer, J.J., 1995 The contexts of Palaeoliths, in A.J. Schofield (ed.), *Lithics in Context*, Lithic Studies Society, 45–51

Wymer, J.J., 1999 *The Lower Palaeolithic Occupation of Britain*, Wessex Archaeology and English Heritage

Xu, S., Anderson, R., Bryant, C., Cook, G.T., Dougans, A., Freeman, S., Naysmith, P., Schnabel, C. and Scott, E.M., 2004 Capabilities of the new SUERC 5MV AMS facility for ^{14}C dating, *Radiocarbon* 46, 59–64

Zagwijn, W.H., 1985 An outline of the Quaternary stratigraphy of the Netherlands, *Geologie en Mijnbouw* 64, 17–24

Appendix 1
Original Flint Artefact Descriptions

Find Number	Type	Description
40/17-12-07/004	Flake	Rather thick centripetal Levallois flake *sensu stricto* (max. dimensions: 104 mm). Some retouch on distal surface and retouch/notch on the lateral face of the artefact shows a different patina. Iron oxide stains from commercial dredging activities. Light grey, fine-grained, flint with characteristic intrusions of lighter coloured dots and spots. Flint shows larger 'circular' coarser-grained zones. At the edges a darker grey zone is more transparent (glassy-like). It has a beige-brown patina and the darker grey zone has light porcelain white to bluish white patina
240/17-12-07/005	Flake	Interpreted as a Levallois flake in a broader sense (max. dimensions: 107.4 mm). Artefact has a dihedral butt but can also be interpreted as a polyhedral butt which is slightly retouched. A grey, fine-grained flint with some lighter coloured stains and dots. At the edge the flint shows a darker grey-bluish (glassy-like) colour. It has a light beige-green (khaki) colour patina with grey bluish coloured zone showing small amounts of bluish-white patina
240/07-12-07/006	Flake	Elongated Levallois flake *sensu stricto*, probably a side scraper (max. dimensions: 150.2 mm). Shows iron oxide stains from commercial dredging activities. Light grey, fine-grained, flint with characteristic intrusions in the form of small lighter coloured dots and spots. Within the flint also more coarse-grained 'circular' inclusions and bigger fossils (crystals up to 13.1 mm) appear. Has an orange-red patina
240/11-01-08/008	Flake	Could also be interpreted as a pseudo (geo-) artefact (max. dimensions: 111.5 mm). One surface seems to be fresher than the other. Light grey, fine-grained flint with characteristic intrusions in the form of small lighter coloured dots and spots. The flint also shows larger coarser-grained zones. At the edges there is a more brown grey, transparent (glassy-like) zone
240/11-01-08/009	Flake	Slightly elongated centripetal Levallois flake *sensu stricto* (max. dimension 107.7 mm). Shows iron oxide stains from commercial dredging activities Light grey, fine-grained flint with characteristic intrusions in the form of small lighter grey dots and spots. At the edges the flint looks more transparent (glassy-like) and has a grey-bluish colour. Orange-red patina. The grey-bluish coloured zone shows porcelain white to bluish-white patina. The more porcelain white-bluish patina seems to be situated around the pressure-cones
240/17-12-07/012	Flake	Thick overstruck flake (max. dimensions: 121.6 mm). Part of the surface looks like a frost fissure (post-flaking). Shows iron oxide stains from commercial dredging activities. Light grey, fine-grained flint with characteristic intrusions in the form of small lighter coloured dots and spots. The flint also shows bigger coarser-grained 'oval' zones. At the edges there is a darker grey transparent (glassy-like) zone with lighter stains. The patina is beige-brown and a darker grey zone shows light porcelain white to bluish-white patina
240/17-12-07/013	Flake	Flake (max. dimensions: 112 mm). A light grey to dark grey, fine-grained flint with characteristic intrusions in the form of small lighter coloured dots and spots. The flint also shows small and lighter 'circular' stains and inclusions (fossils). The patina has a glossy shine with a greasy appearance and light porcelain white to bluish-white. One surface shows more floss patina than the ventral surface
240/17-12-07/014	Flake	Blade-like flake (max. dimensions: 152 mm) and shows iron oxide stains from commercial dredging activities. Light grey, fine-grained flint. At the edges the flint looks darker (grey-green) and is somewhat more glassy-like. A light orange-red patina with an area slightly green-grey patinated. The broken end is bluish-white patina
240/17-01-08/016	Flake	More prepared flake. This flake can be interpreted as a Levallois product in a broader sense (max. dimensions: 110.8mm). The artefact is a tool (3 or 4 big notches), one notch looks less patinated. Light grey, fine-grained, flint with characteristic intrusions in the form of a small lighter coloured dots and spots. The flint also shows larger coarser-grained 'oval' inclusions. At the edges is a black-grey, more transparent (glassy-like) zone. The patina is light brown-beige-brown-greenish colour and the black-grey zone shows mainly a gloss shine with greasy appearance
240/16-01-08/017	Flake	Flake (max. dimensions 146.7 mm). Light grey, fine-grained, flint with characteristic intrusions in the form of small lighter coloured dots and spots. The flint also shows 'white' 'circular' stains and bigger coarser-grained 'oval' zones. At the edges is a dark grey, transparent (glassy-like) zone with lighter stains. The patina is light brown-beige

Find Number	Type	Description
240/17-01-08/018	Flake	Elongated centripetal Levallois flake *sensu stricto*. The bulbar scar has eliminated part of the bulb of percussion force (max. dimensions: 137.7 mm). Light grey, fine-grained flint with characteristic intrusions in the form of a lighter grey and more coarse-grained 'circular' inclusions. The patina is light brown/beige-brown
240/16-01-08/019	Flake	Centripetal Levallois flake *sensu stricto* (max. dimensions: 120.9 mm). One side of the artefacts probably shows some macroscopic use wear and a patinated notch. Grey, fine-grained flint with inclusions in the form of small (lighter) white dots and spots. At the edges the flint looks more transparent (glassy-like) and has a grey-bluish colour. The patina has a grey-bluish coloured zone which shows porcelain white to bluish-white. Mainly gloss sheen with greasy appearance. One surface shows more glass patina than the other
240/17-01-08/021	Flake	Flake (max. dimensions: 114.8 mm). The artefact looks rather fresh. Grey flint with large coarse-grained areas and 'circular' inclusions. At the edges the flint looks more glassy-like. Within the flint fossils to 1–2 mm appear. The patina has a beige-greenish gloss
240/22-01-08/025	Flake	Flake (max. dimensions: 119.8 mm). On one surface are *c.* five notches (which eliminated part of the butt and bulb of percussion/force). These notches are not 'recent' (fresh) but do not show the same patination as the complete artefact (less patinated). Also applies to the scar negative which eliminated the butt. A scar negative on one surface is very recent. Light grey (fine-grained to somewhat more coarser-grained) flint, which shows 'circular' coarse-grained zones. The flint also shows bluish-grey inclusions. The edge has grey-banded colouring. Underneath this grey banded colouring a more bluish coloured 'layer', which is more glassy-like and looks more transparent. Within the flint larger fossil inclusions (up to 7.5 mm) appear. The patina is light orange-red
240/24-01-08/026	Flake	Thick 'elongated' Levallois flake *sensu stricto* (max dimension: 123 mm). One face has probably been exposed to the surface for a longer time. Small parts of the edge are probably retouched. Light grey, fine-grained flint with characteristic intrusions in the form of lighter coloured dots and spots. Some areas have a dark grey colour. The flint also shows coarse-grained 'circular' and 'elongated' inclusions and a grey-brown coloured band. One surface is more patinated than the other. One is light brown to beige-brown colour patina, and the other is light beige/-brown colour
240/17-01-08/032	Flake	Elongated centripetal Levallois flake *sensu stricto* (max dimension: 123.9 mm). One surface has two notches (which eliminated part of the butt and bulb of percussion/force). These notches are not 'recent' (fresh) but do not show the same patina as the complete artefact (less patinated). The artefact shows iron oxide (FeO) stains from commercial dredging activities. Light grey, fine-grained, flint with characteristic intrusions in the form of lighter coloured dots and spots. The flint also shows larger 'circular' darker grey inclusions and somewhat coarser-grained (dark grey) areas. At a surface edge there is a grey-brown banded colouring, which is more glassy-like and looks more transparent. The patina is a green-beige-brownish colour
240/30-01-08/034	Flake	Flake. Can also be interpreted as a biface roughout on a flake (max dimension: 125.4 mm). The retouch on a surface took away the bulb of percussion/force. Light grey, fine-grained, flint with characteristic intrusions in the form of small lighter and darker coloured dots and spots. The flint also shows 'circular' stains (sometimes coarser-grained), inclusions with a white coloured stripe around and small fossils. The patina has mainly a light glossy shine with greasy appearance. The difference with the recent scar negatives is difficult to see
240/30-01-08/035	Flake	More prepared flake. This artefact can be interpreted as a Levallois flake in a broader sense (max dimension: 109.9 mm). Grey, fine-grained flint with characteristic intrusions in the form of lighter coloured dots and spots. The flint also shows larger darker grey inclusions and somewhat coarser-grained areas. At an edge there is a grey-brown banded colouring, which is more glassy-like and looks more transparent. The patina is light brown to beige-brown in colour
240/05-02-08/039	Flake	More prepared flake. This artefact can be interpreted as a Levallois flake in a broader sense (max dimension: 107.9 mm). The flake shows a plain butt, which is also partly facetted. The break is not recent. The artefact shows iron oxide (FeO) stains from commercial dredging activities. Grey, fine-grained, flint with some darker coloured stains. At the edges, the flint shows a darker grey-bluish (glassy-like) colour. The patina is light beige-green (khaki) in colour. The darker grey-bluish coloured zone shows a small amount of porcelain white to bluish-white patina
240/05-02-08/040	Flake	Flake (max dimension: 132.1 mm). White-grey, fine-grained, flint with coarse-grained 'circular' inclusions and some more coarse-grained zones. At right edge surface a lighter grey zone with characteristic intrusions in the form of small lighter coloured dots and spots. Within the flint also fossils up to 2.8 mm appear. The patina is light orange-red in colour
240/05-02-08/041	Flake	Flake (max dimension: 107.2 mm). The artefact looks somewhat rolled. The flake shows iron oxide (FeO) stains from commercial dredging activities. Light grey, fine-grained flint with characteristic intrusions in the form of dark grey dots and spots. The patina is light brown to beige-brown colour

Find Number	Type	Description
240/04-02-08/042	Flake	More prepared flake. This artefact can be interpreted as a Levallois flake in a broader sense (max dimension: 103 mm). The more recent scar negatives show a different glossy shine. Light grey, fine-grained, flint with characteristic intrusions in the form of lighter coloured dots and spots. At the edges a grey-bluish area, which includes light grey 'layers', is described. The patina is a light brown to beige-brown colour
240/04-02-08/043	Flake	Elongated Levallois flake *sensu stricto* (max dimension: 119.2 mm). Light grey, fine-grained flint with characteristic intrusions in the form of small lighter coloured dots and spots. Within the flint also more coarse-grained 'circular' zones and bigger fossils up to 11.7 mm appear. The patina is a light orange-red colour
240/07-02-08/045	Flake	More prepared flake or core? This artefact can be interpreted as a Levallois product in a broader sense (max dimension: 132.9 mm). One surface looks more recent, but shows some colour patina. Probably this flake (or core?) bashed against another rock (see cone and bulb of percussion/force) and split on a frost fissure (the natural fissure is clearly post-flaking). The artefact shows also iron oxide (FeO) stains from commercial dredging activities. Light grey, fine-grained flint with characteristic intrusions in the form of small lighter coloured dots and spots. At the edges there is a darker grey-brown, transparent (glassy-like), zone with intrusions in the form of small lighter coloured dots and spots. The patina is an orange-brown colour patina. The darker grey-brown coloured zone shows light porcelain white to bluish-white patina (at the edge underneath the cortex). The more recent surface of the artefact shows at the edges an orange-brown colour patina (no glossy shine with greasy appearance could be described). This 'ventral' surface consists of a frost fissure which is post-flaking
240/31-01-08/047	Flake	Thick flake (max dimension: 131.8 mm) which can be interpreted as a steep scraper/denticulate (nearly a Quina scraper). If this is correct, than a surface pattern stays centripetal or radial but with 7 scar negatives. Alternatively this artefact can be interpreted as a core. Dark grey, somewhat more transparent (glassy-like), fine-grained flint. At the edges the flint looks more light grey and shows small lighter coloured dots. The patina is orange-red-brown colour. At the edge a brown-grey patina is described. This brown-grey zone shows light porcelain white to bluish-white patina
240/04-02-08/048	Flake	Flake-tool (max dimension: 114.7 mm). Around 20 mm of the flake edge (on the ventral surface) is retouched. Light grey, fine-grained, flint with characteristic intrusions in the form of small lighter coloured dots and spots. The flint also shows more coarse-grained 'circular' inclusions. The patina is light brown to beige-brown-green colour
240/12-02-08/049	Flake	Thick hinged flake (max dimension: 149 mm). The artefact shows pseudo cortex and iron oxide (FeO) stains from commercial dredging activities. Light grey, fine-grained flint with characteristic intrusions in the form of small lighter grey dots. The patina is light brown/beige-brown colour. On a surface some beige-brown 'oxidation' stains are described
240/13-02-08/050	Flake	Flake (max dimension: 115.8 mm). Light grey, fine-grained flint with characteristic intrusions in the form of lighter grey dots and spots and somewhat more coarse-grained 'circular' inclusions. At the edges, underneath the cortex, the flint looks more transparent (glassy-like) and has a grey-bluish colour. The patina is light brown/beige-brown-greenish colour. The grey-bluish coloured zone shows porcelain white to bluish-white patina
240/12-02-08/054	Flake	Flake (maximum dimension = 117.1 mm). The recent and lateral (patinated) edges are broken. The artefact shows iron oxide (FeO) stains from commercial dredging activities. Light grey-grey, fine-grained flint with characteristic intrusions in the form of small lighter coloured dots and spots. Within the flint also more coarse-grained zones and 'circular' inclusions appear. Some darker coloured dots and spots are noticed as well. The flint contains bigger 'circular' fossils up to 11.9 mm. The patina is light brown/beige-brown colour
240/12-02-08/055	Flake	Interpreted as an *éclat débordant* (max dimension: 110.6 mm). Light grey, fine-grained flint with characteristic intrusions in the form of small lighter coloured dots and spots. The flint also shows larger coarser-grained 'oval' inclusions. At the edges dark grey, transparent (glassy-like), zone is described. The patina is orange-brown colour patina. The dark grey, transparent (glassy-like), coloured zone shows a light porcelain white to bluish-white patina at the edge
24011-02-08/056	Flake	More prepared flake. This artefact can be interpreted as a Levallois flake *sensu stricto* (max dimension: 104.9 mm). The artefact is a scraper with possible Quina retouch. Dark grey, fine-grained glassy-like, flint with characteristic intrusions in the form of lighter coloured dots and spots. The patina is porcelain white-bluish. This porcelain white-bluish patina seems to be situated around the pressure-cones

Find Number	Type	Description
240/07-02-08/058	Flake	Flake (max dimension: 107.1 mm). In the middle of a surface there seems to be some possible macroscopic use wear (20 mm). Grey, fine-grained, flint with characteristic intrusions in the form of small lighter coloured dots and spots (more stained). Within the flint also more coarse-grained zones appear. At the edges, the flint shows a darker grey-bluish colour. The patina is light brown/beige-green colour. The darker grey-bluish coloured zone shows a small amount of porcelain white to bluish-white patina
240/18-03-08/060	Flake	Centripetal 'elongated' Levallois flake *sensu stricto* (max dimension: 122.1 mm). The artefact is also interpreted as a single convex side scraper with surface retouch. Some sediment residue, present on the artefact, was sent for analysis to the *Rijksdienst voor Archeologie, Cultuurlandschapp en Monumenten* (*RACM*). The tool was also briefly examined for microscopic use wear traces by Mms Dr Veerle Rots (University of Leuven, Belgium). The artefact shows iron oxide (FeO) stains from commercial dredging activities. Light grey, fine-grained, flint with characteristic intrusions in the form of small lighter coloured (white) dots and spots. Within the flint also more darker grey (coarse-grained?) stains are present. At the edge, the flint looks more glassy-like and has a darker grey colour. The patina is light beige-green/beige-grey colour
240/18-03-08/061	Flake	More prepared flake. This artefact can be interpreted as a Levallois flake in a broader sense (max dimension: 100.1 mm). The artefact shows iron oxide (FeO) stains from commercial dredging activities. Light grey/grey, fine-grained flint with characteristic intrusions in the form of small lighter dots and spots. The patina is light brown/beige-brown colour
240/03-03-08/065	Flake	Flake (max dimension: 87.6 mm). Grey, fine-grained flint with characteristic intrusions in the form of small lighter coloured dots and spots. The flint also shows larger coarser-grained 'oval' zones/inclusions. At the edges a lighter grey and fine-grained zone with characteristic intrusions in the form of small lighter coloured dots and spots are described. The patina is light brown-beige colour. The more coarser-grained ('oval' zones/inclusions) are beige/grey patinated
240/25-02-08/066	Flake	Flake (max dimension: 111.5 mm) which shows on both cortex. The artefact is heavily rolled. Black/dark grey, fine-grained, flint with light grey coarse-grained inclusions (up to 23 mm). The patina is orange-brown-grey colour. The grey-brown patinated zone at the edges shows light porcelain white to bluish-white patina
240/13-02-08/067	Flake	More prepared flake. This artefact can be interpreted as a Levallois flake in a broader sense (max dimension: 101.8 mm). The distal (recent) and lateral (recent) parts are broken. The artefact shows iron oxide (FeO) stains from commercial dredging activities. Grey, fine-grained flint with characteristic intrusions in the form of small lighter coloured dots and spots (looks more stained with coarse-grained zones). At the edges, the flint looks more transparent/glassy-like and has a darker yellow/beige colour. The latter zone has a little bit bluish-white patina. The patina is light beige-green (khaki) colour patina. The grey-bluish coloured zone shows a little bluish-white patina
240/18-03-08/068	Flake	More prepared flake. This artefact can be interpreted as a Levallois flake in a broader sense (max dimension: 111 mm). Distal (recent) and lateral (recent) parts are broken. The artefact shows iron oxide (FeO) stains from commercial dredging activities. Dark grey, fine-grained flint with characteristic intrusions in the form of small lighter coloured dots and bigger lighter coloured stains. At the edges the flint looks grey-brown and is somewhat more transparent (glassy-like). The latter zone has a more porcelain white to bluish-white patina (mainly on the ventral surface). The patina is light glossy patina with porcelain white to bluish-white patinated zones (mainly on the ventral surface, around the bulb of percussion/force)
240/03-03-08/069	Flake	Flake, which could also be a pseudo (geo-) artefact (max dimension: 112 mm). On a surface possibly a notch is present (could be recent?).The different scar negatives show different degrees of patination. Light grey to dark grey, fine-grained, flint with characteristic intrusions in the form of small lighter coloured dots and spots. The flint also shows larger coarser-grained 'oval' zones. The dark grey zone seems to be more concentrated at the edges and is more transparent (glassy-like). The patina shows a light glossy shine with greasy appearance. There is a clear difference in patina between the different scar negatives. Part of a frost fissure (pre-flaking) near the butt, shows light porcelain white to bluish-white patina. A later 'frost fissure' on the dorsal surface, cut by several scar negatives, is slightly orange-brown patinated. On this dorsal surface several white chalky concretions are visible
240/14-03-08/070	Flake	Thick 'secondary' flake which is clearly produced on a fluvial transported pebble (max dimension: 130.9 mm). The artefact shows iron oxide (FeO) stains from commercial dredging activities. Light grey, fine-grained flint with characteristic intrusions in the form of dark grey coloured dots and spots. The flint also shows 'circular' inclusions. At the (right) edge of the ventral surface, underneath the cortex, the flint looks darker (dark grey/black) and is more glassy-like. The patina is light orange-red colour

Find Number	Type	Description
240/15-02-08/071	Flake	Flake with possibly some retouch at the right lateral edge (max dimension = 114 mm). The artefact shows iron oxide (FeO) stains from commercial dredging activities. Light grey, fine-grained flint with characteristic intrusions in the form of lighter grey dots and spots. Some zones look more coarse-grained (oval-like inclusions) than others. At the edges, underneath the cortex, the flint appears more transparent (glassy-like) and has a grey-bluefish colour and a lighter coloured band. Orange-red colour patina. The grey-bluefish coloured zone shows porcelain white to bluish-white patina. The more porcelain white-bluish patina seems to be situated around the pressure-cones
240/03-03-08/072	Flake	Large elongated flake (max dimension = 172 mm). It is difficult to distinguish older and more recent scar negatives (mainly on the dorsal surface) from each other. Dark grey to black, fine-grained, flint. The flint also shows smaller and larger light grey, but more coarse-grained, 'circular' inclusions and some lighter coloured areas. A surface shows a light brown/orange/beige-brown colour patina. The other surface shows a more grey-brown patina
240/14-03-08/073	Flake	Flake, which could also be a pseudo (geo-) artefact (max dimension = 116.5 mm). The ventral surface shows one 'old' scar negative (patinated notch?). The flake shows iron oxide (FeO) stains from commercial dredging activities. Light grey, fine-grained, flint with characteristic intrusions in the form of lighter grey and dark grey dots and spots. At the edges, underneath the cortex, the flint looks darker (grey-black) and somewhat glassy-like. The ventral surface shows a light beige-brown colour patina. The dorsal surface shows a darker grey-green colour patina. The dorsal surface is more patinated than the ventral surface
240/14-03-08/075	Flake	Flake (max dimension = 106.1 mm). On both edges there is a recent break. Light grey, fine-grained flint with characteristic intrusions in the form of small lighter coloured dots and spots. Within the flint also more coarse-grained 'circular' zones appear. At the edges the cortex, the flint looks more transparent (glassy-like) and has a darker grey-bluish colour. The latter zone has porcelain white to bluish-white patina. Also bigger fossils (crystals) up to 14.1 mm appear. Light brown/beige-green colour patina. The darker grey-bluish coloured zone shows porcelain white to bluish-white patina. The more porcelain white patina seems to be situated around the pressure-cones. The dorsal surface seems to show more gloss patina than the ventral surface
240/07-12-07/081	Flake	More prepared flake. This artefact can be interpreted as an elongated Levallois flake in a broader sense (max dimension: 161 mm). Part of the butt is 'missing' (shattered) on the dorsal surface (part of the cone is still visible). Dark grey, fine-grained, flint with characteristic intrusions in the form of lighter coloured dots and spots. Some areas have a lighter grey colour. The flint also shows coarse-grained 'circular' inclusions. Dark green (dark khaki) colour patina
240/14-03-08/082	Flake	Flake (max dimension: 88.8 mm). The artefact shows frost fissures (post-flaking). At the edge of the frost fissure the patina looks somewhat thicker. Probably some frost cracks were already present in the artefact before it was recovered. During dredging activities/transport the artefact fell eventually apart. A 'recent' scar negative on the dorsal surface is younger than the frost fissure. Light grey, fine-grained to somewhat coarse-grained flint with characteristic intrusions in the form of lighter grey dots and spots. Some parts of the flint are more coarse-grained. At the edges, the flint looks more transparent (glassy-like) and has a grey-bluish colour. Light brown/beige-brown colour patina. The grey-bluish coloured zone shows porcelain white to bluish-white patina
240/15-02-08/086	Flake	Slightly overstruck flake (max dimension: 104.3 mm). Possibly the bulb of percussion/force is missing due to retouch? Light grey, fine-grained, flint with characteristic intrusions in the form of small darker coloured dots and spots. At the edges there is a dark grey, more transparent (glassy-like), zone. Glossy shine with greasy appearance and light porcelain white to bluish-white patina on the darker grey, more transparent (glassy-like), zone
240/17-01-08/022	Core	The artefact can be interpreted as a Levallois *sensu stricto* core (nucléus Levallois à éclat préférentiel) which was split by frost actions (max dimension: 110.1 mm). The core edge periphery measures 333 mm. One side completely represents a frost fissure, which is post-flaking. The other side shows a centripetal pattern. Light grey, fine-grained flint with characteristic intrusions in the form of small lighter coloured dots and spots. The flint also shows 'white' and somewhat darker stains and bigger coarser-grained 'oval' zones. At the edges (underneath the cortex) a dark grey, transparent (glassy-like), zone with lighter stains is described. Beige-brown-orange colour patina. The dark grey zone shows light porcelain white to bluish-white patina
240/31-01-08/033	Core	More prepared core which can be interpreted as an elongated Levallois core in a broader sense. The core produced blade-like flakes (max dimension: 145.9 mm). The striking platform on the core is retouched; some preparation. The core edge periphery measures 371 mm. Light grey, fine-grained, flint with lighter and darker coloured zones. At the edges there is a darker brown, more transparent (glassy-like), zone (*c.* 5 mm thickness). Beige-brown-green (khaki) colour patina

Find Number	Type	Description
240/25-01-08/038	Core	Probably a disc core with a centripetal pattern (max dimension: 123.7 mm). The striking surface shows 12 scar negatives. There is cortex on the dorsal face. The core edge periphery measures 370 mm. Light grey/green to darker grey, fine-grained flint with characteristic intrusions in the form of small lighter coloured dots and spots. The flint also shows larger coarser-grained zones. Orange-brown colour patina. The larger coarser-grained zones show a grey colour patina. The striking surface looks less patinated (colour- and gloss-patina) than the flaking surface
240/11-02-08/046	Core	Core on a large cortex covered flake (flaked-flake, max dimension: 113.5 mm). A surface of the flake shows 7 secondary negatives which form a 'parallel' + lateral unidirectional pattern. The core edge periphery measures 321 mm. Light grey/green to darker grey, fine-grained flint with characteristic intrusions in the form of small lighter coloured dots and spots. The flint also shows bigger coarser-grained zones. Porcelain white to bluish-white patina
240/15-02-08/062	Core	Probably a disc core with a centripetal pattern on one side (max dimension: 100.5 mm). A side shows a total of 13 scar negatives. The cortex (cortex and pseudo cortex) is situated on the other side. The core edge periphery measures 291 mm. Light beige/grey, fine-grained flint with characteristic intrusions in the form of small lighter coloured dots and spots. The flint shows underneath the cortex a green (khaki-coloured) to grey zone with black stains. Directly underneath the cortex a thin black line is visible. Orange-brown colour patina
240/28-02-08/063	Core	The artefact can be interpreted as a Levallois *sensu stricto* core (nucléus Levallois récurrent) which produced more than one preferential flake (max dimension: 100.2 mm). A side shows 17 scar negatives and the core edge periphery measures 300 mm. Light grey, fine-grained flint with characteristic intrusions in the form of small lighter coloured dots and spots and white inclusions. At the edges there is a dark grey, transparent (glassy-like), zone with lighter stains. Light brown-green (dark khaki) colour patina
240/29-02-08/064	Core	Probably a disc core with a 'parallel' bi-directional pattern (max dimension: 136.1 mm). One side looks 'more recent', but with colour patina. Probably this core (or flake?) bashed during, fluvial transport, against another rock and split on a frost fissure (frost fissure is post-flaking). The artefact shows iron oxide (FeO) stains from commercial dredging activities. The core edge periphery measures 429 mm. Light grey, fine-grained, flint with characteristic intrusions in the form of small lighter coloured dots and spots. The flint also shows bigger coarser-grained zones. At the edges there is a darker grey, transparent (glassy-like), zone with lighter coloured stains. Beige-brown colour patina. The darker grey zone shows light porcelain white to bluish-white patina
240/18-03-08/074	Core	Disc/discoidal core on a frost split piece of flint (max dimension: 147.3 mm). On the centre of a side a cluster of several points of impact/force (around 5) is noticed. The artefact could be interpreted as an anvil or hammerstone. A total of 6 scar negatives are described on this side). The core edge periphery measures 415 mm. Dark grey, fine-grained flint with characteristic intrusions in the form of small lighter coloured dots and spots. The flint also shows larger coarser-grained 'oval' zones, a laminated area and some smaller fossils. Dark brown-green (khaki) colour patina
240/17-12-07/001	Hand axe	Cordiform hand axe (max length = 117.3 mm). One side shows a number of hinged scar negatives. One side is worked more intensively than the other. Grey, fine-grained, flint with characteristic intrusions (and inclusions) in the form of lighter grey dots and spots. The flint also shows somewhat coarser-grained, brown-beige zones (inclusions). Light grey/beige colour patina
240/07-12-07/002	Hand axe	Sub-triangular hand axe (max length: 121.9 mm). One side shows a number of hinged and stepped scar negatives. This side is worked more intensively than the other side. Light grey, fine-grained flint with characteristic intrusions (and inclusions) in the form of lighter grey and white dots and spots. The flint also shows somewhat coarser-grained areas. Light grey/beige colour patina. Mainly gloss patina
240/17-12-07/003	Hand axe	Tip of a broken hand axe which is considered flat (max length = 122.7 mm). Light grey, fine-grained, flint with characteristic intrusions (and inclusions) in the form of lighter coloured (grey) dots and spots. The flint also shows somewhat coarser-grained areas. Light grey/beige colour patina, with lighter orange stains
240/04-01-08/007	Hand axe	Cordiform hand axe (max length = 130.2 mm). One side is worked more intensively than the other side. One side shows a number of hinged and stepped scar negatives. The hand axe is probably produced on a large and thick flake. The artefact shows iron oxide (FeO) stains from commercial dredging activities. Grey-beige, fine-grained, flint with characteristic intrusions in the form of lighter coloured dots and spots. The flint also shows coarser-grained areas. At the edges there is a more transparent (glassy-like) grey-blue zone. Light beige colour patina. The grey-blue coloured zone on one side shows a bluish-white patina

Find Number	Type	Description
240/11-01-08/010	Hand axe	Between cordiform and ovate hand axe (max length: 122.7 mm). Initially the flake was overstruck and secondarily the bulb of percussion was eliminated by 'retouching'. One side (and part of the former butt, near the cortex) shows some natural fissures. Light grey, fine-grained flint with characteristic intrusions in the form of small lighter coloured dots and spots. The flint also shows somewhat more coarse-grained 'circular' inclusions and fossils. At the edges there is a more transparent (glassy-like) zone of approx. 11 mm thickness. The latter zone has a grey-bluish colour. Light brown/beige-brown/greenish colour patina
240/11-01-08/011	Hand axe	Part of a broken hand axe which is considered flat. Could be a roughout of a hand axe on a flake (max length: 127.7 mm). The plain area beside the cortex on one side can be interpreted as the former butt. The flake was produced by means of hard percussion. Light grey, fine-grained flint with characteristic intrusions in the form of small lighter coloured dots and spots. The flint also shows somewhat more coarse-grained 'circular' inclusions. At the edges is a more transparent (glassy-like) zone of approx. 12 mm thickness. The latter zone has a grey-bluish colour. Light brown/greenish-grey colour patina
240/17-01-08/015	Hand axe	Cordiform hand axe (max length: 117.4 mm). One side is worked more intensively than the other side. One side shows some a number of hinged scar negatives. Light grey, fine-grained flint with characteristic intrusions in the form of lighter coloured (light grey to white) dots, spots and inclusions. The flint also shows grey 'oval', somewhat coarse-grained, zones. Light grey/beige colour patina (mainly on one side). Mainly gloss patination. At the edges there is a bluish-white patina
240/17-01-08/020	Hand axe	Sub-triangular hand axe (max length: 129.7 mm). One side shows a single hinged scar negative. Light grey, fine-grained flint with characteristic intrusions in the form of lighter coloured (white) dots, spots and darker coloured stripes/zones. At the edge there is a more transparent (glassy-like) grey zone. Light grey/beige colour patina. Mainly gloss patination and some areas with bluish-white patina
240/17-01-08/023	Hand axe	Tip of a broken hand axe which is considered flat (max length: 80.2 mm). One side shows a number of hinged scar negatives. The other shows some natural fissures. Light grey, fine-grained flint with characteristic intrusions in the form of small lighter coloured dots and spots. The flint also shows some more coarse-grained 'circular' inclusions. At the edges there is a more transparent (glassy-like) zone. The latter zone has a grey-bluish colour. Green/beige/brown colour patina
240/24-01-08/024	Hand axe	Base of a broken hand axe which is considered flat (max length: 110.1 mm). One side shows a number of hinged and stepped scar negatives. The artefact shows Bryozoa on both sides of the piece. Bryozoa are also known as moss animals, sea mats, polyzoa, corallines and ectoprocts. They usually occur in colonies in the shallows. Grey, fine-grained flint with characteristic intrusions in the form of lighter coloured (white) dots and spots. The flint also shows coarser-grained areas and inclusions. At the edges there is a more transparent (glassy-like), grey-blue zone. The patina is bluish-white
240/22-01-08/027	Hand axe	Sub-cordiform hand axe (max length: 137.3 mm). One side is worked more intensively than the other side. The other side shows a single hinged scar negative. The hand axe shows some cortex on its base. A crushed zone can be seen on one side. Light grey to darker grey, fine-grained flint with lighter coloured (white) zones. The flint also shows coarse-grained inclusions and coarse-grained 'circular' fossils (approx. 15 mm in diameter). Light beige to grey colour patina. Mainly gloss patination and some bluish-white patina at the edges
240/22-01-08/028	Hand axe	Sub-cordiform hand axe, but mainly a thick scraper on one side (max length: 102.4 mm). This bifacial tool is probably produced on a large and thick flake. One side is worked more intensively than the other side. This side shows a number of slightly hinged scar negatives. The artefact shows iron oxide (FeO) stains from commercial dredging activities. Light grey, fine-grained flint with characteristic intrusions (and inclusions) in the form of lighter coloured (grey) dots and spots. Some of these dots and spots have a reddish-brown edge. The flint also shows somewhat coarser-grained areas. Light beige colour patina. Mainly gloss patination
240/21-01-08/029	Hand axe	Probably a cordiform hand axe (max length: 125.5 mm). The tip of the hand axe is missing. Both sides show a hinged scar negative. Light grey, fine-grained flint with characteristic intrusions in the form of small lighter coloured dots and spots. The flint also shows somewhat more coarse-grained 'circular' inclusions. On the outside of the artefact there is a more 'tea-stain like (brown) coloured layer' (this is probably a colour patination). Brown-beige ('flamed') colour patina with black stains
240/21-01-08/030	Hand axe	Cordiform hand axe (max length: 140.7 mm). Light grey to darker grey, fine-grained flint with characteristic intrusions in the form of lighter coloured (white) dots and spots. The flint also shows grey coarser-grained areas and inclusions. At the edges there is a more transparent (glassy-like), grey-blue, zone/stripe. Light beige/brown colour patina. Mainly gloss patination. The grey-blue zone on one side shows a bluish-white patina

Find Number	Type	Description
240/21-01-08/031	Hand axe	Base of a broken hand axe which is considered flat, probably a cordiform hand axe (max length: 117.2 mm). One side is worked more intensively than the other. On one side there is a brown/orange iron oxide like (FeO) stain. The base of the hand axe shows two straight sides on which a number of impact points. Light grey, fine-grained flint with characteristic intrusions in the form of lighter coloured (light grey to white) dots, spots and inclusions. The flint also shows somewhat coarse-grained grey zones. Light grey/beige colour patina. Mainly gloss patination. A bluish-white patina is situated at the edge of one side
240/24-01-08/036	Hand axe	Sub-cordiform hand axe (max length: 124.3 mm). Both sides show a number of hinged and stepped scar negatives. One side shows some natural fissures. Light grey, fine-grained, flint with characteristic intrusions in the form of small lighter coloured dots and spots. The flint also shows 'circular', somewhat coarse-grained, inclusions and fossils. Underneath the cortex, there is a more grey-bluish zone. Green/beige/brown colour patina
240/25-01-08/037	Hand axe	Elongated (thick) sub-cordiform hand axe. Could be interpreted as a roughout of a hand axe on a flake (max length: 126.7 mm). The plain area beside the cortex can be interpreted as the butt. A side shows some natural fissures. Light grey to darker grey, fine-grained flint with lighter coloured (grey) zones. The flint also shows coarse-grained 'circular' inclusions. The cortex is around 8 mm thick. Light beige to grey colour patina. The colour patina is mainly situated on one side
240/04-02-08/044	Hand axe	Part of a broken hand axe which is considered flat (max length: 108.3 mm). Dark grey, fine-grained flint with characteristic intrusions in the form of lighter grey dots and spots (zones). Bluish-white patina (mainly on one side)
240/13-02-08/051	Hand axe	Tip of a broken hand axe which is considered flat (max length: 135.2 mm). Both sides show a number of hinged and stepped scar negatives. Light grey, fine-grained flint with characteristic intrusions in the form of lighter coloured (white) dots and spots. The flint also shows darker coloured zones and fossils. Brown-orange-beige colour patina on one side. Brown-beige colour patina on the other side. Bluish-white patina is concentrated around the cortex
240/13-02-08/052	Hand axe	Sub-cordiform hand axe on a flake (max length: 98.3 mm). One side is worked more intensively than the other side. Probably a big notch is situated on one side. This side also shows a number of hinged and stepped scar negatives (in the vicinity of the base). Light grey to darker grey, fine-grained flint with characteristic intrusions in the form of lighter coloured (light grey and white) dots and spots and inclusions. The flint also shows a white (stripe-like), somewhat coarse-grained, zone. Light grey/green colour patina. Mainly gloss patination. A bluish-white patina is situated at the most extreme outside of the artefact
240/12-02-08/053	Hand axe	Part of a broken hand axe which is considered flat (max length: 86.2 mm the maximum length and width, without orientating the artefact, are respectively 102.4 mm and 88.0 mm). One side shows a number of hinged and stepped scar negatives. The tip of the hand axe is broken anciently (the braking surface shows gloss patination). Darker grey, fine-grained flint with characteristic intrusions in the form of lighter coloured (light grey to white) dots and spots and inclusions. The flint also shows grey (stripe-like), somewhat coarse-grained zones. Light grey colour patina. Mainly gloss patination. A bluish-white patina is situated at the most extreme outside of the artefact
240/07-02-08/057	Hand axe	Between *amygdaloid à talon* and *ovalaire à talon* hand axe. Could be interpreted as a roughout of a hand axe (max length: 109.5 mm). Both sides show several hinged scar negatives. One side shows some natural fissures. Probably only hard percussion was used to produce this artefact. Light grey, fine-grained flint with characteristic intrusions in the form of lighter coloured dots and spots and inclusions. The flint also shows darker, somewhat coarse-grained, zones. At the edges there is a more transparent (glassy-like) zone. The latter zone has grey-bluish colour. Green-beige-brown colour patina. The grey-bluish coloured zone has porcelain white to bluish-white patina
240/15-02-08/059	Hand axe	Part of a broken hand axe which is considered flat (max length: 98.8 mm). One side is worked more intensively than the other side. The hand axe is probably produced on a flake. One side shows a number of hinged and stepped scar negatives. Part of the base is broken off recently. The artefact shows iron oxide (FeO) stains from commercial dredging activities. Light grey, fine-grained, flint with characteristic intrusions in the form of lighter coloured (white) dots and spots. The flint also shows grey coarser-grained areas which are surrounded by a darker grey coloured flint. Light beige colour patina (mainly on one side). Mainly gloss patination. One negative is patinated bluish-white
240/31-01-08/076	Hand axe	Cordiform hand axe? sub-triangular hand axe (max length: 142.7 mm). Both sides show a number of hinged and stepped scar negatives. The tip of the hand axe is missing (recent break). Light grey, fine-grained flint with characteristic intrusions in the form of small lighter coloured dots and spots. The flint also shows 'circular', somewhat coarse-grained, inclusions and a more transparent grey-bluish zone. Green-beige colour patina

Find Number	Type	Description
240/22-01-08/077	Hand axe	Between *amygdaloid à talon* and *ovalaire à talon* hand axe (max length: 118.3 mm). One side is thicker than the other side. Both sides show a hinged scar negative. On the thickest point of one side some pressure cones are described. Light grey/beige, fine-grained flint with coarser-grained areas and inclusions. At the outside of the artefact there is a brown-beige 'layer' of 2.3 mm (probably patina?). Brown-orange colour patina. Directly underneath the patina, the flint looks more abraded/weathered (brown-beige 'layer' of around 2.3 mm)
240/14-01-08/078	Hand axe	Sub-cordiform hand axe (max length: 122.8 mm). Both sides show a hinged scar negative. Light grey, fine-grained flint with characteristic intrusions in the form of lighter coloured (white) dots and spots. The flint also shows 'circular' coarse-grained inclusions. At the edge there is a more transparent (glassy-like), grey-brown, zone. Light beige/brown colour patina. The grey-brown zone (underneath the cortex) shows a bluish-white patina
240/17-12-07/079	Hand axe	Cordiform hand axe (max length: 138.2 mm). The complete base of the hand axe is cortex covered. The artefact also shows pseudo cortex. One side shows some hinged scar negative. The other side has some natural fissures. Dark grey, fine-grained, flint with characteristic intrusions in the form of small lighter coloured dots and spots. The flint also shows somewhat more coarse-grained 'circular' and 'elongated' inclusions. At the edges there is a more transparent (glassy-like) zone, which has a grey-bluish colour. Green-beige-brown colour patina. A porcelain white to bluish-white patina is situated at the edges
240/29-01-08/080	Hand axe	Sub-cordiform hand axe (max length: 132.8 mm). One side shows a single hinged scar negative. This side is worked more intensively than the other side. Light grey, fine-grained flint with characteristic intrusions in the form of small lighter coloured dots and spots. The flint also shows somewhat more coarse-grained 'circular' and 'elongated' inclusions. At the edges there is a more transparent (glassy-like) zone of approx. 12 mm thickness. The latter zone has a grey-bluish colour. Green-beige-brown colour patina. The grey-bluish colour zone shows porcelain white to bluish-white patina
240/14-03-08/085	Hand axe	*Discoide à talon* hand axe (max length: 117.8 mm). Probably a more evolved roughout of a hand axe. One side is worked more intensively than the other side. Both sides show several hinged scar negatives. One side shows some natural fissures. The measurement of the complete edge periphery (including the butt) is 342 mm. Light grey, fine-grained flint with characteristic intrusions in the form of small lighter coloured dots and spots. The flint also shows somewhat more coarse-grained 'circular' and 'elongated' inclusions. At the edges (underneath the cortex) there is a more transparent (glassy-like) zone, which has a grey-bluefish colour. Green-beige-brown colour patina. The grey-bluish colour zone shows porcelain white to bluish-white patina
240/..-..-../087	Hand axe	Artefact not available for study
240/..-..-../088	Hand axe	Artefact not available for study
240/18-03-08/089	Hand axe	Cordiform hand axe (max length: 102.5 mm). One side shows some hinged scar negative. On the base of the hand axe cortex is described. Both sides of the artefact show natural fissures. The artefact shows Bryozoa on both sides of the piece. Bryozoa are also known as moss animals, sea mats, polyzoa, corallines and ectoprocts. They usually occur in colonies in the shallows. The artefact also shows iron oxide (FeO) stains from commercial dredging activities. Grey-greenish, fine-grained flint with characteristic intrusions in the form of small lighter coloured (white) dots and spots. The flint also has grey 'oval' coarse-grained zones. At the edge there is a more darker grey zone. Light beige colour patina. Mainly gloss patination and some bluish-white patina at the edges
240/07-02-08/090	Hand axe	(Sub) cordiform hand axe (max length: 117.7 mm). One side shows some hinged scar negative. Both sides show cortex remains. Light grey to darker grey, fine-grained flint with characteristic intrusions in the form of lighter coloured (white) dots and spots. The flint also shows darker stripe of approx. 21 mm thickness and coarse grained inclusions and fossils. Brown-beige colour patina on both sides. One side seems to possess more colour patina. A bluish-white patina is situated at the edges of the artefact

Index

by Susan M. Vaughan

Page numbers in *italics* denote illustrations.

Abbeville (France) 57
Aggregates Levy Sustainability Fund 6–8, 102
Ancaster, River 39
animal bones
    Area 240
        assemblage 17, 19, 95, 102, 103
        dredger monitoring 92, 93
        geological context 103–4
        sampling survey 83–5, *85*, 86
    chronological discussion 45–6, 51, 56, 57, 61
    submerged sites 73, *74*, 101–2
Archaeological Exclusion Zone 1, 2, 40, *40*, 86, 88, 94, 95, 108
Archaeological Recovery Zone 11, 28, 40, *40*, 95, 96
Arco Adur 87, 88, *88*
Area 240
    assemblage formation and post-depositional modification 95
        artefact assemblage 95–6
        geological context 96–7
        raw material and production 98
        site formation scenarios 98–100, *99*
        summary 100
        taphonomy 97–8, *97*
    assessment of dredge loads 88–9, *89*
    discovery 1–2, *2*
    discussion 104–6, 111
    dredger monitoring 86–7
        assessment of loads 88–9
        flint 89–92, *91*
        palaeontological and environmental material 92–3
        sampling strategy 88, *88*
        at SBV Flushing Wharf 89, *89*
        summary 94
    geographic and cultural setting 100–3, *101*
    management and mitigation 108–9
    method evaluation 106–8
    original assemblage
        animal bones 17
        discussion 17–19
        flint 11–17, *13*, *14*, *15*, *16*
        source 11, *11*
    project background 2–3
    project contribution to field of submerged prehistory 8–9
    project funding 6–8
    sampling strategy *87*, 83, *88*
    seabed sampling survey 78
        discussion 85–6, 94
        grab sample acquisition and processing 81–5, *81*, *82*, *83*
        positioning 78
        strategy 78, *79*
        two metre scientific trawl 80–1, *81*
        video/stills photography 78–80, *79*, *80*
    study area 1, *1*, 3–5, *3*, *4*
    see also palaeogeographic reconstruction
Area 251 66–8
Area 254 60
Area 360 101–2
Area 464 8
Arun, River 8, 25, 28, 75–7, *76*
Audit of Current State of Knowledge of Submerged Palaeolandscapes and Sites 9
aurochs (*Bos primigenius*) 56, 57, 103

Baltic 32
Baltic/Eridanos River 48
Barley Picle 50
bear 61
Beeches Pit (Suffolk) 50
bison (*Bison priscus*) 17, 56, 102, 103
Bobbitshole (Suffolk) 56
Bouldnor Cliff (Solent) 22, 73, 108
Boxgrove (W Sussex) 18, 51, 104, 105
Breydon Formation 21, 41, 42–3, *42*, 65, 66, 68
Breydon Water (Norfolk) 41, 49
British Marine Aggregate Producers 6
Britons Lane Formation 52
Broom (Devon) 19
Broome Terrace 42, 70
Brown Bank Formation 58, 60, 62, 103
Bure, River 50, 65
Bytham (Ingham), River 39, 45–6, 48, 49, 70

Caister-on-Sea (Norfolk) 51
Caister Road 50
Caister Shoal 43
Cefas Endeavour 25
chalk rafts 49
charcoal 33–5, 56, 66
Cooperation in Science and Technology 8
cores, flint
    assemblage 12–14, *13*, *14*, 17–18, 95
    descriptions 129–30

taphonomy 97–8
Corton Till Member 48
Coston (Norfolk) 57
Crag Group 45, 46
Cromer Forest-bed Formation 45, 46
Cromerian Complex 45–6, 48
Cross Sand 43, 48–9
Crown Estate 4, 5, 11

Dartford (Kent) 58
Denmark, submerged sites 73
diatom assessment 33, 35
diving, problems associated with 73–4
Doggerland 109
Dunbridge (Hants) 107–8

East Coast Regional Environmental Characterisation
    survey 7, 25–7, 38, 55, 96, 107
    flint 77–8, *77*
East English Channel survey 75, 109, 110
Ebbsfleet (Kent) 18, 51
Eem Formation 57
Elbow Formation 103
Elveden (Suffolk) 18
English Heritage 5, 6, 7, 9, 86, 87
Environmental Impact Assessment 5–6, 8, 74, 106, 109

fallow deer (*Dama dama*) 56
Fermanville (France) 73
fire, use of 50
flakes, flint
    assemblage 12–14, *13*, 17–18
    context 95, 96, 97
    descriptions 125–9
    dredger monitoring 90, 92
    sampling survey 83, *83*, 85
    site formation scenarios 98–100, *99*
    taphonomy 97–8, *97*
flint
    Area 240
        assemblage 11–12, 95–6
        characterisation 12–15, *13, 14, 15, 16*
        descriptions 125–33
        discussion 17–19
        dredger monitoring 89–92, *91*
        geological context 96–7
        post-depositional modifications 16–17
        raw material 15–16, 98
        sampling survey 83, *83*, 84, 85, 86
        significance 102–3
        site formation scenarios 98–100, *99*
        taphonomy 97–8, *97*
    Area 360 101–2
    Arun survey 75–7
    chronological discussion 45, 50–1, 52, 61, 62, 68
    East Coast REC project 77–8, *77*
        river systems 39–40

from submerged sites 73, *74*
    see also cores; flakes; hand axes
flint knapping 12, *12*, 50, 95–6
foraminifera analysis 33, 35
    Pre-Anglian 46
    Saalian 56
    Devensian 58
    Holocene 65, 66
Frindsbury Wharf (Kent) 93

giant deer (*Megaloceros dawkinsi/giganteus*) 56, 102, 103
Goat's Hole (Glam) 61
Gray Mammoth 78
Great Witchingham (Norfolk), Lenwade Pits 39
Great Yarmouth (Norfolk) 41
Guidance Note 6

hand axes
    assemblage 11, *13*, 14–15, *15*, 16, 17–19
    context 95, 96, 97, 104
    descriptions 130–3
    dredger monitoring 89–92, *91*
    site formation scenarios 98–100, *99*
    taphonomy 97–8
Happisburgh (Norfolk), survey 8, 31, 39, 45, 104, 105, 109
    flint 18
Happisburgh Formation 48
Harnham (Wilts) 19
High Lodge (Suffolk) 18, 104
hippopotamus (*Hippopotamus amphibius*) 56
Homersfield Terrace 42, 52, 70
hominin/human activity 104–6, 109–10
    Pre-Anglian 45
    Hoxnian 50–1
    Saalian 52, 102
    Ipswichian 56–7
    Devensian 58, 61–2
    early Holocene 62, 65, 68
Homo antecessor 109–10
Homo heidelbergensis 51, 109
Homo neanderthalensis 61, 104, 105, 109, 110
Homo sapiens 61, 109
horse (*Equus bressanus/caballus*) 17, 102, 103
Hoxne (Suffolk) 18, 39, 50–1, 100
human activity *see* hominin/human activity
Humber survey 75
Hundred Stream 50
hyena 61

Ingham, River *see* Bytham, River
insect remains 33–5
Irish elk *see* giant deer

Kessingland (Suffolk) 68
Keswick Mill Pit (Norfolk) 39, 100, 102
kettle holes 51

lion (*Panthera leo*) 56, 61
Lion Tramway Cutting (Essex) 18–19
Lowestoft (Suffolk) 41, 45
Lowestoft Till Formation 48, 51
Lynford Quarry (Norfolk) 19, 39, 61

magnetometer system 31, *31*
mammoth (*Mammuthus* sp) 17, 61, 92, 95, 102, 103, 105
marine aggregate dredging 3–6
Marine Aggregates and Archaeology: a Golden Harvest 7
Marine Aggregates Levy Sustainability Fund 6–7, 102
Marine Environment Protection Fund 7–8
Marine Management Organisation 5
Meulmeester, Jan 11
Meuse, River 45, 61
Mineral Industry Research Organisation 7
molluscan analysis 33–5, 56, 58, 65, 68
moose (*Alces latifrons*) 102, 103
Muelmeester, Jan 1

Neanderthals *see Homo neanderthalensis*
Neumark-Nord (Germany) 57
Newarp Banks 50
North Sea Palaeolandscapes Project 7–8, *7*, 22
North Sea Prehistory Research and Management Framework 8
Norton Subcourse (Norfolk) 39
Norwich (Norfolk), Carrow Road 39, 100

Optical Stimulated Luminescence dating
    channel and floodplain deposits 46, 55, 56, 58–60, 62, 96
    evaluation 106
    samples 25, 35, 36
ostracod analysis 33, 35, 56, 58, 65, 66

Pakefield (Suffolk), survey 8, 31, 39, *41*, 45, 104, 105, 109
    flint 18
palaeogeographic reconstruction
    methods 21–2
        evaluation 106–8
        existing data review 22–8, *23, 24, 25, 26, 27*
        geophysical data acquisition 28–32, *28, 29, 30, 31*
        palaeoenvironmental assessment, analysis and dating 33–6, *34, 35*
        palaeoenvironmental sampling 32–3, *32, 33*
        seabed sampling 32
        subsequent work 36, *37*
        summary 36–8
    Palaeo-Yare and Area 240 39–41, *40*, 43–5, *44*
        Pre-Anglian (MIS 13 upwards) 45–8, *46, 47*
        Anglian (MIS 12) 48–9
        Late Anglian (MIS 12) 49–50, *49, 50, 51*
        Hoxnian (MIS 11) 50–1, *51*
        Saalian (MIS 10–6) 51–6, *51, 52, 53, 54*
        Ipswichian (MIS 5e) 56–7, *56*
        Devensian (MIS 5d–MIS 2) 57–62, *57, 59, 60, 63*
        early Holocene and final transgression 62–8, *64, 67*
        post-transgression development 68–9, *69*
        sediment preservation 69–71
peat 65, 66–8, 93, 96, 102
People and the Sea: A Maritime Archaeological Research Agenda for England 8
photography 78–80, *79, 80*
plant remains 33–5, 56, 58, 60, 66, 68
pollen analysis 33, 35
    Pre-Anglian 46
    Saalian 55, 56
    Devensian 58, 60, 62
    Holocene 66
Pontnewydd Cave (Clwyd) 19, 105
Portchester (Hants) 18
positioning 29, 78
prehistoric characterisation 39–41, *40*; *see also* palaeogeographical reconstruction
Protocol for Reporting Finds of Archaeological Interest 1, 6, 73, 95, 101, 103, 109
Purfleet (Essex) 52

radiocarbon dating
    Holocene 64, 65, 66, 68
    method evaluation 106
    sampling 33–5, 35–6
    summary 17
red deer (*Cervus elaphus*) 56, 102, 103
Regional Environmental Characterisation surveys 7–8, *7*, 22
reindeer (*Rangifer tarandus*) 17, 103
resources 104–5, 106
Rhine, River 45, 48, 61
rhinoceros (*Stephanorhinus hemitoechus*) 56, 57; *see also* woolly rhinoceros
Rotterdam (Neths) 107
Runham (Norfolk) 41

sandbanks 43, 68
SBV Flushing (Neths) 1, *1*, 2, 11, 12, 87, 89, 95
Scheldt, River 61
Scroby Sand 43
sea-level change 4–5, *4*, 8, 104, 105
    Pre-Anglian 45
    Anglian 48, 69
    Hoxnian 50, 51
    Saalian 51, 52, 56, 70
    Ipswichian 56, 57, 70
    Devensian 57–8, 60, 61, 62
    early Holocene 65
Seabed Prehistory: Gauging the Effects of Marine Aggregate Dredging 7, 22, 24, 25, 55
seabed sampling 32, 73–4

evaluation 107–8
grab sampling 75–8, *75*
remote techniques 74–5
see also under Area 240
Sheringham Cliff Formation 52
sidescan sonar 30, *30*, 31
single-beam echosounder 29
spearheads, bone 73
straight-tusked elephant (*Palaeoloxodon antiquus*) 56, 57, 102
Strijdonk, Henk 1
sub-bottom profilers 24–5, *24*, *28*, 29–30, *29*, 31, 32
submerged forests 21
Submerged Prehistoric Landscapes and Archaeology of the Continental Shelf 8
Swanscombe (Kent) 18, 51
swath bathymetry 25, *25*, 28, 29, 32

Thames, River 39, 45, 48, 50, 61
Thurne, River 50
Twente Formation 61

Uphill (Som) 61

Valdoe Quarry (W Sussex) 105
vibrocore survey 32–3, *32*, *33*
Vlissingen (Neths) 103

Warren Hill (Suffolk) 104
Waveney, River 5, 41–2, 49, 50, 52, 57, 62, 68
Wensum, River 5, 39, 41, 49
Wessex Explorer 28
West Coast Palaeolandscapes Project 7
West Runton (Norfolk) 45, 104
Westkapelle Ground Formation 46
Whitlingham (Norfolk) 39, 50, 100
Wissey, River 61
wood fragments 56, 68, 92–3, 102
woolly rhinoceros (*Coelodonta antiquitatis*) 17, 61, 102, 103
Wortwell (Norfolk) 42, 57

Yare, River, Palaeo-Yare Valley
    assessment 36, 38
    geographic and cultural setting of assemblage
        assemblage significance 100, *101*, 102–3
        potential for material in aggregate block 100–2
    onshore research 38
    palaeogeographical reconstruction *see* palaeogeographical reconstruction
    present-day setting *40*, 41–3
Yare Valley Formation 21, 41–2, 52, 58, 65
Yarmouth Roads Formation 45, 46, 48, 85, 103

Zeeland Ridges 61